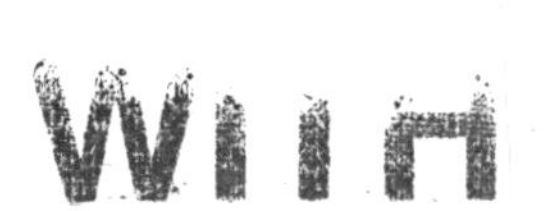

D1084535

Beaks, Bones, and Bird Songs

BEAKS, BONES, AND BIRD SONGS

How the Struggle for Survival Has Shaped Birds and Their Behavior

ROGER J. LEDERER

TIMBER PRESS • PORTLAND, OREGON

Frontispiece: Some Blue Jays migrate but others are sedentary, preferring to face the rigors of the winter environment.

Credits for photos and illustrations appear on pages 263–266.

Published in 2016 by Timber Press, Inc.
The Haseltine Building
133 S.W. Second Avenue, Suite 450
Portland, Oregon 97204-3527
timberpress.com

Printed in China
Text design by Adrianna Sutton
Jacket design by Skye McNeill

Library of Congress Cataloging-in-Publication Data

Names: Lederer, Roger J., author.
Title: Beaks, bones, and bird songs : how the struggle for survival has shaped birds and their behavior / Roger J. Lederer.
Description: Portland, Oregon: Timber Press, 2016. | Includes bibliographical references and index.
Identifiers: LCCN 2015045261 (print) | LCCN 2015051354 (ebook) | ISBN 9781604696486 (hardcover) | ISBN 9781604697520 (e-book)
Subjects: LCSH: Birds—Evolution. | Birds—Behavior, Classification: LCC QL677.3 .L43 2016 (print) | LCC QL677.3 (ebook) | DDC 598—dc23
LC record available at http://lccn.loc.gov/2015045261

A catalog record for this book is also available from the British Library

CONTENTS

Teachers have influenced me throughout my entire life. I have worked with teachers at all levels of education. I have been a teacher. With enormous admiration and respect I dedicate this book to all those who spend their lives in such a noble profession, educating others about the natural world and how we are part of it.

FOREWORD

I'm a birder, so I'm biased, but I think birds are the most wonderfully diverse, unfailingly fascinating creatures on the planet.

With more than 10,000 species, birds encompass almost every possibility of shape, lifestyle, niche, design, and environment. There are hummingbirds scarcely bigger than bees, and flightless cassowaries that stand as tall as a human, with a kick that can kill a person (though they rarely do). Birds of paradise shimmer with color and cascades of gauzy plumage; nightjars blend so well with the dead leaves of the forest floor that one must almost step on them to see them. For the benefit of his mate, a male Great Argus—a pheasant from Southeast Asia—fans his immense, specialized wing feathers and long tail to create a mesmerizing half circle of silver, from which hundreds of iridescent golden eye spots seem to glare.

Birds are found from the North Pole to the fringes of Antarctica, from steamy Bornean rainforests to the driest deserts on Earth. Seabirds crisscross the vastness of the world's oceans effortlessly; some albatrosses, in the half decade or more before they are old enough to breed, may clock 100,000 miles in a single year, wandering the gale-raked seas of the Southern Hemisphere.

Birds eclipse even the highest mountains on the planet, flying over—not around—the Himalayas at nearly 30,000 feet. Imagine sitting on the wing of an airplane instead of inside the pressurized cabin, in subzero temperatures, gasping for breath in the low-oxygen atmosphere—and now imagine exerting yourself like a sprinter running a four-minute mile. That's what some cranes do twice a year.

I've been privileged to study wild birds for many years, exploring the lives and movements of some of the most interesting groups. For two decades,

I've led a study of the migration of the Northern Saw-whet Owl, a tiny raptor that weighs about as much as a robin, and is so infrequently seen that many birders go a lifetime without encountering one. When we started studying these owls in the mountains of Pennsylvania, Saw-whets were thought to be so rare that they were the symbol of the state wildlife conservation fund. But by stretching our whisper-thin nets in the woods at night each autumn, and playing the weird, tooting call of the male Saw-whet, we catch them by the hundreds—or more. One remarkable season, we netted almost 4000 of them, traveling south from the boreal forests of Canada.

Another group I've been working with for years are hummingbirds, which add zip and color to backyards from the Rio Grande to Alaska. Science has peeled back many of the mysteries of these minute dynamos—their extraordinary physiological abilities, their epic migrations, their ferocious metabolisms—but we still have a lot to learn. I've worked with colleagues across the continent to explore the way that some western hummingbird species, most notably the Rufous Hummingbird of the Pacific Northwest, are rapidly evolving new migratory routes and new wintering areas in the East and Southeast. Each winter—yes, winter!—we catch and band dozens of vagrant hummingbirds, which happily tolerate temperatures well below freezing in the middle of northeastern winters. One hardy Rufous Hummingbird has been documented surviving nighttime temperatures down to -9°F, with windchills below -30°F. That's amazing for any animal, much less one that weighs slightly more than a penny. How do they survive? These hummingbirds have the ability to drastically lower their body temperature at night, essentially going into a nightly hibernation to conserve energy, then emerging from torpor at first light, and resuming their hunt for tree sap, winter-active insects, and other food.

All this is by way of whetting your appetite for what you'll find in this book. Educator, birder, and ornithologist Roger Lederer is your guide through the myriad, and often almost miraculous, ways that birds live their varied lives. From avian senses to communication, flight to physiology, migration to sex—the whole range of bird life and natural history is here.

You'll learn how birds react to hurricanes; that goldfinches double the density of their plumage for winter; why their small size makes thieves of

Calliope Hummingbirds; and how (in a sense) urban birds understand posted speed limits for automobiles, perhaps better than human drivers.

I've had the pleasure of birding with Roger on his home ground in the mountains and river valleys of California. Under his entertaining tutelage, you'll never look at birds the same way again. And you may see that it is not bias, but a simple fact: birds really are the most fascinating animals on Earth.

—Scott Weidensaul, author of *Living on the Wind* and other books

INTRODUCTION

It's Tough To Be a Bird

There are some four million different kinds of animals and plants in the world.
Four million different solutions to the problems of staying alive.

—DAVID ATTENBOROUGH

Look outside your window, take a walk, go fishing, watch a video, or do anything else that allows you to see birds in the wild. You may get the impression that birds are blithely going about their business, happily chirping, singing, and scratching among the leaves, flitting from branch to branch, clambering up a tree trunk, or soaring through the sky barely moving a feather. Looks like an easy life. Cultural symbols like the dove representing peace, the bluebird signifying happiness, and the robin as the harbinger of spring reinforce the idea that birds have not a care in the world. But we don't often see the arduous challenges a bird faces every moment of every day.

Of the many hours I have spent in the field, watching birds flying, feeding, resting, and nesting, I was most affected by those moments when I saw birds searching for food in blowing snow, sitting on the surface of an ocean fighting threatening waves, and flying in serious winds. I wondered: how do birds make it from hatching to adulthood and from year to year after that?

Birds have to be on task all the time. They have to use their senses to find food, migrate, withstand the weather, avoid predators, compete with each other and alien species, and face a myriad of other trials. This book is about the abilities, adaptations, and behaviors birds possess and employ to survive from one day to the next. It is only the most physiologically, anatomically,

The phrase "free as a bird" implies a carefree existence and the liberty to go anywhere, anytime, but birds are not as free as their aerial life implies.

and behaviorally well-tuned birds who successfully meet these challenges and go on to the most important goal in their life, reproduction.

Accurate figures for mortality and longevity of wild birds are nearly impossible to determine, but there are trends. Only about 50 percent of White-eyed Vireos in the southeastern United States return to their breeding grounds from their winter quarters, and merely 36 percent of Downy Woodpeckers, resident all year throughout much of North America, survive from one year to the next. Songbird adults have a 40–60 percent survival rate from year to year. Of their young, perhaps only 10 percent make it from egg to adulthood the following year. This means that a two-year-old songbird is a one out of twenty miracle. This short life expectancy is a result of the many dangers birds face. And while bigger birds have a lower mortality rate than smaller birds, they all face hazards every day. Unlike humans, birds don't seem to get closer to their demise as they age; instead of slowly declining, most birds, after reaching maturity, have an equal chance of dying suddenly at all times of their precarious life. Sick or injured birds are rarely seen in the wild as illness or injury puts them at immediate risk of death, so you only see healthy birds on your bird walks.

Evolution has been at work on birds for more than 200 million years, shaping them into adept and adroit organisms. But birds today not only face the challenges that natural selection throws at them, but an entirely new set of obstacles, thanks to us. Things started to change for birds shortly after humans came on the scene. Early humans incorporated birds into their diet. Then agriculture came along, usurping habitat but also inadvertently providing food for birds. As civilization matured, bird feathers, bills, and bones became adornments for human culture; later, birds were domesticated for meat or eggs. Bird hunting became more efficient with the advent of guns and as civilization spread, habitat shrank. Massive destruction of wildlife of all sorts was common, including the commercial hunting of many bird species for food and feathers, until the passage of the Lacey Act in the United States in 1900—the first federal law protecting wildlife. The Migratory Bird Treaty Act of 1918 further protected migratory birds. These laws made a difference in North America, but many bird species live in or migrate through areas of the world that don't pay much attention to the needs of birds or their protection. Today about 1400 of the world's 10,000 bird species are threatened with extinction.

We have destroyed habitats and replaced them with cities, highways, tall windowed buildings, transmission towers and lines, microwave antennas, wind turbines, and lights—along with millions of cats. Birds have always contracted diseases, but humans have altered the environment and allowed pathogens to spread more quickly. A solar plant in the Mojave Desert concentrates the sun's rays so strongly that birds are incinerated if they happen to fly through it. Birds have evolved rather amazing and often unique adaptations to the environment, but as the world changed, those adaptations became increasingly less effective. Birds never evolved defenses against windows or lights, buildings or towers, or the large number of our feline friends. Climate change has caused changes in bird migration patterns, but what the long-term effect will be is unknown.

With ornithological science as the background, this book will explore the common and unusual ways birds put into operation their physical and behavioral adaptations. What everyday challenges does a bird face and how does it survive? Seeing ultraviolet, finding food without seeing or touching it, flying thousands of miles nonstop, maneuvering deftly and speedily through thick forests, navigating by smell, surviving extremes of weather, sharing community resources, and changing their songs in noisy cities are just some of the amazing things birds do to simply make it to tomorrow and cope with the challenges of a changing planet.

BIRDS, BEAKS, AND BELLIES

The Whys and Wherefores of Foraging

In birds the mouth consists of what is called the beak, which in them is a substitute for lips and teeth. This beak presents variations in harmony with the functions and protective purposes which it serves.

—ARISTOTLE, *On the Parts of Animals*

Ever watch birds jockeying for position to snatch the best morsels of seed at a bird feeder? The same phenomenon occurs in the woods, but is not as easily observed as it happens in a much larger venue with a wider variety of birds and food items. Foraging, from the Old French *fourrage* (to forage, pillage, or plunder), refers to the ways birds find food. This activity, followed by feeding, consumes much of a bird's day. And for good reason: efficient foraging is indispensable for survival. Birds have many other challenges (weather, predators, competitors, migratory journeys) on the way to their fundamental goal of reproduction, but these only add to the burden of finding food.

The feeding behavior of birds has been studied for many years, but in the 1960s ornithologists recognized that birds that maximize their energy intake per unit time spent foraging produce the most offspring. Those birds that ate foodstuffs with less nutritional value or spent too much—or not enough—time seeking food are no longer with us. As a result of this revelation, the importance of foraging in avian ecology has been reflected in a large percentage of field studies ever since.

Successful foraging is the result of beak (or bill, the terms are interchangeable) shape, which in turn determines much of a bird's lifestyle. The beak, with few exceptions, is the bird's only tool. Birds use their beak not only to forage and feed, but also to preen and oil feathers, defend territories, attack predators, build nests, and aid in courtship displays. So if it is an all-around tool, why aren't all bird bills the same—some perfect all-purpose beak created by evolution? Because natural selection, in its wise ways, reduced competition by making different bills for different foods.

THE EVOLUTION FROM TEETH TO BEAK, THE ULTIMATE TOOL

"Rare as hen's teeth" has come to mean rarity itself. Of all the vertebrate groups (fish, amphibians, reptiles, birds, and mammals) only birds lack teeth as we know them: enamel outcroppings fixed in jaws. About 150 million years ago in the Jurassic Period there lived an almost perfect example of the transition between reptiles and birds—an animal with rooted teeth, a long bony tail, and the abdominal ribs of reptiles, but also feathers. This animal, *Archeopteryx* (ancient wing), is certainly among the most important fossils ever found and for many years was considered to be the first bird. However, since its discovery in 1861, many other fossils of potential bird ancestors have been uncovered, such as the Chinese fossil *Xiaotingia zhengi*, discovered in 2011.

Many early birds shared major skeletal characteristics with coelurosaurian (hollow-tailed lizard) dinosaurs: forwardly located pelvic (hip) bones, large bony eye sockets, light, hollow bones, reduced tail vertebrae, elongated arms and hands, and fused clavicles (making the wishbone). They also shared a similar egg structure and some had feathers. As birds evolved, improvements in flying ability required morphological changes, such as the need to weigh less. Teeth became smaller and reduced in number until they finally disappeared. Lighter bills, both hooked and serrated, replaced heavier teeth. Along with bills, birds developed other ways to grasp prey such as

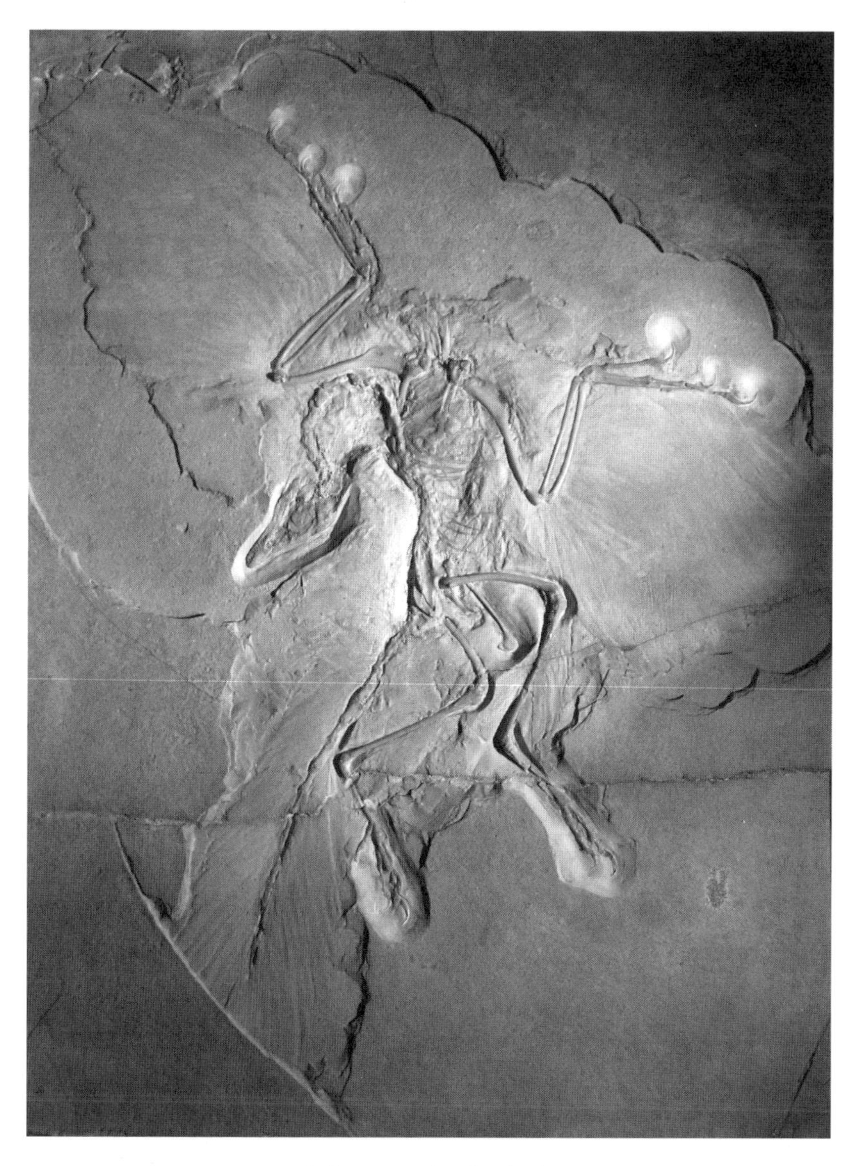

Archeopteryx, the "first bird."

tongues and feet with spines or prickles. Whereas toothed mammals chew food to begin digestion, birds simply snatch and quickly swallow food like insects, nectar, fruit, worms, and seeds. To accommodate these changes, the feeding mechanisms of birds became specialized not only in the form of beaks but further down, bellies (crops and gizzards).

A bird's beak is composed of an upper and lower bony jaw covered by a thick layer of keratin, a structural protein, the same substance that forms skin, feathers, scales, fingernails, and turtle shells. This is the rhamphotheca (Greek for beak case or sheath), which grows throughout the year, sometimes changing color seasonally as it does in the European Starling whose wintery black bill becomes yellow in the spring.

Beaks, which evolved for the purpose of food getting, are as varied in shape and size as ice cream flavors, reflecting the diversity of food and its source. Beaks range in size from the African Shoebill's enormous wooden shoe–shaped beak to the miniscule beaks of small finches. Hooked, long, thick, wide, pointed, blunt, up- or downturned, bent, crossed, swollen, or serrated, beaks tear, probe, suck, filter, chip, crack, tweeze, chisel, crush, strain, spear, or seize food items.

Along with differing shapes, styles, and lengths of beaks came different feeding styles. Heavy conical bills handle seeds well, whereas flat triangular bills catch flies easily. Crossbills use their overlapping mandibles for opening pine cones; long-billed shorebirds probe deeply into mudflats locating their quarry without seeing, touching, or smelling it; and skimmers slice through the top of the oceans' waters, snatching food items near the surface. Every bird has its own beak-defined niche; a flycatcher could not survive by feeding on mudflats any more than a sandpiper could live in thick woods.

Bills often serve as sexual signals as well. Male Zebra Finches with bright bills attract females as do puffins with their multicolored beaks. Bird bills also help to regulate body temperature by radiating heat, partially compensating for birds' lack of sweat glands. So sometimes there is a bit of a compromise among the functions of the bill, but eating is the highest priority.

Consider the finches of the Galapagos Islands and their bill shapes. An ancestral finch or two landed on the Galapagos from South or Central America. The birds increased in number and spread to different islands until eventually 14 different species with beak variations came to be. Three species of finches eat seeds off the ground, three more live on cactuses and eat mainly fruit and insects, and one seedeater lives in trees. The rest are arboreal insect eaters, including the tool-using Woodpecker Finch that extracts insect larvae

Several Galapagos (Darwin's) Finches showing a diversity of bill shapes.
1. *Geospiza magnirostris* 2. *Geospiza fortis* 3. *Geospiza parvula* 4. *Certhidea olivacea*

from tree branches with the use of a cactus spine. Evolving a different beak meant exploiting a new food source and sharing the food supply, which benefited everyone.

Every bird has its own technique of seeking out and acquiring sustenance with beak shape defining a range of behaviors and food items. Since we are most familiar with the birds that visit our feeders, let's start with the seedeaters.

SEEDEATERS OR GRANIVORES

Granivory (seed and grain eating) evolved in tandem with birds developing the ability to fly as seeds provide accessible sources of concentrated energy. The lower jaws of granivores are solidly muscularized, enabling the jaw to push the seed upward into the almost immoveable upper jaw. The hard palate of the upper jaw is heavily keratinized and has ridges, bumps, and spine-like projections that serve to husk the seed and direct the digestible part backward.

Seeds are not easy to digest, so after the seeds are swallowed they move to the crop (from the Old English *cropp*, meaning craw), an expanded part of the esophagus. Most bird species have a crop, but some, like owls and geese, do not. The crop may have evolved as part of the active lifestyle of some birds. Fossil evidence from China indicates that perhaps as far back as 140 million years ago some birds had crops for temporary food storage as did some herbivorous dinosaurs. Rather than eating food slowly and digesting it before moving on, birds with a crop fill it to almost bursting. Digestion then begins as the food slowly makes its way downward. I once found a dead grouse with a fist-sized crop full of juniper needles. Veterinarian Thomas Caceci dissected a Wood Duck and found 10 decent-sized acorns in its crop; that would be like a human throat filled with 10 golf balls.

From the crop the food goes to the small intestine and then the glandular part of the stomach that secretes digestive enzymes, the proventriculus (before the small belly). The second part of the stomach is the ventriculus (small belly) or gizzard, from the Old French *gésier*, chicken entrails. The gizzard is muscular and, substituting for teeth, mechanically grinds food; it often contains sand grains or small rocks to help in the process. Apparently this organ has fascinated people for years. Spallanzani, an Italian Catholic priest of the 18th century, claimed that he fed turkeys scalpel blades, which the gizzard ground to pieces. Some say that if you hold a live chicken up to your ear you can hear gizzard stones grinding.

From the basic conical shape, seedeater bills vary to match the contours of seeds and their different sizes, shapes, and hardness. Deeper bills have more musculature and exert greater forces, so in general, the deeper the bill, the larger the seed that can be handled. Evening Grosbeaks—so named by French explorers who thought the birds only fed in the evening—have a large bill that can exert enough force to crack open cherry pits. Following robins and other cherry eaters that digest the fruit and regurgitate the seeds, Evening Grosbeaks feast on the pits. The European Hawfinch can crack open olive and plum pits as well and feeds its young by regurgitating partially digested seeds.

Birds typically choose the seeds that are most available and easily handled. At a bird feeder you might find White-crowned Sparrows gingerly manipulating seeds to husk them, Eurasian Collared Doves gulping larger seeds, and tiny goldfinches picking through the seed pile for smaller, softer seeds. Finches place the seed laterally on the edge of their lower jaw and slide their jaw slightly forward and back to crack it; to husk it, the bird moves the seed to the middle of the palate and moves its jaws laterally until the shell comes off. Watch the dining habits of these birds at your feeder closely and you'll discover lots of different styles, like your relatives at Thanksgiving dinner.

The Red Crossbill is a specialist. Its crossed mandibles and strong jaws enable the bird to pry open the scales of pine cones and extract the small seeds—the bird holds a cone with one foot while extracting a seed. (Interestingly, individuals with the lower mandible crossed to the right hold the cone with their right foot, and left-crossed birds use their left foot.) This adaptation gives crossbills almost exclusive access to this particular seed source, but specialization has its downside. As seed removal takes time, crossbills need seed abundances two to three times greater than other bird species to fill their daily energy requirements as they have less time to forage. The production of seeds in coniferous forests goes down every three to five years, at which time the crossbills are at a disadvantage competing for other kinds of seeds that they can't handle as well.

Male Red Crossbill; the females are greenish in color. The lower mandible crosses either to the left or right, half of the population being lefties and the other half righties.

Acorn Woodpeckers of the western United States and Mexico store acorns in "granary" trees and defend them aggressively. They wedge the acorns into holes in trees or wooden telephone poles so tightly that crows, squirrels, and rats can't raid their supply. To remove an acorn, a woodpecker hammers it with its bill to crack the shell and extract the meat. Clark's Nutcrackers, capable of carrying more than 90 pine seeds at a time in a pouch under their tongue, store many of them in caches, even under the snow. They cache two to three times what they need for the winter and eventually find half or more of their seed caches later. Not only do the birds recall the site of these caches for up to nine months, they also remember the relative number of seeds and the size of the seeds in each cache. Florida Scrub Jays cache food by burying one acorn at a time; if they observe another jay, a potential cache robber, watching them, they will return later to move the acorn. But they will only do this if they themselves were cache robbers in the past. Seems that honest jays trust the other ones and thieves do not.

THE GRAZERS: BIG, SMALL, FLYING, AND FLIGHTLESS

Lots of birds are grazers, and some of them are considered crop pests—blackbirds in North America and Java Sparrows in Indonesia eat rice crops, while parrots damage almond crops in Australia. The Red-billed Quelea of Africa may be one of the worst pests, because, some say, it is the most abundant bird in the world; super-colonies of an estimated 30 million birds have been observed. Flocks are so large that when they land in trees they break large limbs off. A large flock of quelea can eat 50 tons of grain a day and since quelea are kept as pets in Australia, Queensland Biosecurity is concerned about their possible escape and damage to corn, wheat, and cereal crops. One quelea adaptation for survival is the behavior of breaking up into small search parties to hunt for food and then returning to the colony to transmit information about the new food source. Studies indicate, perhaps not surprisingly, that the number of birds and variety of bird species that feed on food crops is much higher in organically grown fields than in non-organic ones. The numbers of insects, as well as weedy plants, are also double or triple in organic crop fields because of the absence of pesticides.

While fishing on my favorite lake, I admired the numerous Canada (not Canadian) Geese overhead, on the water, and in the shoreline grass. Once in serious decline in the early 20th century because of overhunting and habitat destruction, their current North American population may be nearly six million. The birds graze on grass blades, stems, and seeds, grasping the plants with the lamellae (sharp ridges) of their bills. Not possessing a crop, they eat constantly and the not-so-nutritious food, with a large amount of minimally digestible cellulose, moves quickly through the digestive system. This results in a lot of bird pooping, about every 20 minutes, a big reason the birds are considered pests on school grounds, parks, and golf courses. The flightless Kakapo or owl parrot from New Zealand is also a dedicated vegetarian, regurgitating indigestible fiber. It has a small gizzard for a plant eater, probably because its jaw, tongue, and beak structure allow it to grind up plant matter before swallowing. Unusual among land birds, the Kakapo can also store a large amount of body fat, making it the world's heaviest parrot. It is also the world's rarest parrot and perhaps the longest-lived bird at an estimated 90 years.

Grouse and ptarmigan digest about a fifth of the fiber they eat. If they had evolved a fermentation chamber as part of their digestive system, they would be able to process more, but that would add weight to these birds, which are already weak fliers. Adding a fermentation chamber allowed the evolution of flightless birds such as the Ostrich, Emu, and rheas. These large ground dwellers graze on green plants and seeds, digesting much of the cellulose they ingest by fermenting it in their caecum, comparable to our appendix. The efficient gut of the Emu has a muscular gizzard with a strong grinding ability, aided by grit (in addition to small stones in the gizzard, pieces of glass, wood, and metal are occasionally found) and an acid environment. These birds prefer high-energy foods such as fruit and seeds and are able to extract enough energy from plant stems to support up to two-thirds of their daily energy needs. Ostriches, whose gizzards might contain three pounds of material, have been known to eat rings, bottle caps, spark plugs, bicycle valves, and even pieces of baling wire. However, they do not eat tin cans or hide their heads in the sand.

GLEANERS, HAWKERS, AND PROBERS: THREE WAYS TO EAT BUGS

Another major food source is bugs, really arthropods, of every stripe—spiders, flies, millipedes, ants, beetles, and relatives. Birds catch these nutritious creatures in three ways: gleaning, hawking, or probing.

Gleaners

Gleaning is a type of foraging strategy in which birds pick bugs off the ground, leaves, rocks, or tree trunks. Many gleaning birds, such as tits and kinglets, flutter, hop, hang, or hover glean in the foliage, as they pluck at their prey. Warblers make their way through a tree, nipping bugs off leaves, while thrushes and towhees move along the ground scratching for prey items among the litter. Verdins of the southwestern United States and Mexico use their strong feet to pull leaves toward them or hang upside down to inspect the undersides for tiny prey. They will also eat a larger food item such as a caterpillar by grasping it under one foot and consuming it piece by piece. Black-capped Chickadees don't just move around randomly hoping to find insects. They find caterpillar prey by searching for leaves damaged by caterpillars. They look for dry curled leaves or tree branches and inspect them, distinguishing between caterpillars that have fed on trees with toxic or distasteful substances, such as tannins and glycosides, and palatable prey. After finding a caterpillar, Black-capped Chickadees will often hang upside down by one foot while pecking at the larva held in the other foot.

Although the gleaning group is generally composed of small insectivorous birds, many other birds glean—woodpeckers, quail, grackles, crows, robins, gulls, pigeons, and turkeys. Some have unusual food sources. Oxpeckers of Africa glean parasites like ticks off the back of various ungulates like giraffes and rhinos. The assumption for many years was that this was a case of mutualism—the birds got food and the hoofed animals were rid of skin parasites. However, a new study determined that only 15 percent of the time that birds

Mountain Chickadee demonstrating its agility by hanging on a sunflower head.

spent on the backs of ungulates is devoted to devouring ticks; the rest of the time the birds fed on skin wounds, ear wax, and other goodies found by probing through the hair. Cattle Egrets follow in the footsteps of buffalo, wildebeest, zebra, or cattle, feeding on the invertebrates the mammals scare up. It is clearly profitable as the cattle-following egrets spend two-thirds the amount of energy and get up to three times as much food in a given amount of time as do egrets that forage in the absence of cattle. The birds tend to follow ungulates that walk at a moderate pace because a slower speed does not stir up enough insects and a faster pace hurries their feeding.

Other birds have more indirect methods of gleaning insects. House Sparrows have learned the trick of waiting in a highway diner's parking lot to glean the freshly killed insects off the grills and radiators of cars. Boat-tailed Grackles have been observed doing the same thing in the parking lot of the Kennedy Space Center in Florida. Great-tailed Grackles survive in Death Valley, California, by picking insects off the license plates of tourists' cars; the grills and radiators are presumably too hot.

Hawkers

Hawking birds fly out from a perch, snatch a bug, and then return to the branch or post. Also called sallying or flycatching, this technique is a good way of making a living, as evidenced by hundreds of species of birds. The flycatcher family, Tyrannidae, is the most diverse family of birds with more than 400 species, including the smallest songbirds in the world (such as the pygmy-tyrants) and the species with the longest tail relative to body size of any bird (the Fork-tailed Flycatcher). The Black Phoebe, a common flycatcher resident of the southwestern United States and Central America, sits on a low branch near a creek or pond and waits for a flying insect such as a bee, wasp, or beetle to approach. After flying out and snapping its hooked-tip bill down on the bug, the bird will eat it in mid-air, or, if the bug is big, take it back to the perch and whack it on the branch a few times to kill it. Flycatchers, like many birds in temperate regions, choose their prey by size: the largest flycatchers eat the largest insects, the smallest flycatcher the smallest, and the medium-sized birds take intermediate-size prey.

My PhD thesis was a study of the foraging habits and diet of seven species of flycatchers in different habitats. After doing the requisite library research, I discovered that most ornithologists thought that flycatchers swept insects into their bills with their long rictal bristles, modified feathers on either side of the jaw. But because no evidence supported this idea, I captured flycatchers, put them in a large flight cage, and filmed the birds at 400 frames per second as they caught flies in mid-air. Viewed at normal speed, the films showed the birds snapping up flies in the tips of their rapidly closing bills, with the help of a small downward hook on the upper bill; rictal bristles played no direct role. Later anatomical studies indicated that these bristles have sensory connections to the brain that probably help the bird determine speed and orientation in flight.

Swifts are successful and abundant around the world, numbering about 100 species. With their thin, knife-like wings emanating from a cigar-shaped body and their rudimentary tails, swifts resemble a boomerang. They lead an aerial lifestyle, feeding, drinking, mating, and even sleeping on the wing. They rarely stop except to build a nest. Their feet are so

Flycatcher showing rictal bristles.

small that they cannot perch but instead cling to vertical walls when they do rest (they were once lumped with the hummingbirds into the order Apodiformes; "apodi" means without feet). Normally feeding at moderate heights, swifts may go as high as 3000 feet to feed. While raising young, a pair of White-throated Swifts might bring as many as 5000 arthropods to their nestlings each day.

Swallows, which have somewhat wider wings and longer tails, are aerial feeders although they will occasionally swoop down to pluck insects off the ground or the surface of a lake. They tend to fly fairly close to the ground and choose larger insects than swifts but avoid stinging ones like wasps and bees. Swallows perch regularly on wires or tree branches to rest and occasionally land on the ground although they walk only awkwardly. In times of insect scarcity, swallows will partake of fruit. The Greater Striped Swallow of Africa even eats acacia seeds and feeds them to its nestlings.

The similarly shaped but larger nighthawks and relatives are crepuscular feeders, venturing out in the evening when flying insects are abundant. Their bills are small, barely noticeable, but their open jaws reveal a huge sticky mouth. Nighthawks and relatives are sometimes referred to as goatsuckers because their wide mouth was once thought to allow the birds to suckle on the teats of goats. Flying low and slow with seemingly erratic wingbeats,

Tree Swallows—so named because they nest in tree cavities or nest boxes—winter farther north than any other swallow species.

nighthawks scoop insects into their open jaws and rapidly send them down the gullet, catching a few thousand insects in an evening.

The honeyguides of Africa (family Indicatoridae) is an unusual group of insect eaters, although they are actually omnivorous. They guide humans, such as the Boran tribe of East Africa, to bee colonies. The Boran people whistle to the birds to attract them and the birds respond by signaling the humans with chattering sounds, flying a short distance, and calling again—until the hive is reached. The humans take the honey and the birds get the bee eggs, larvae, pupae, and beeswax. Everybody eats.

Probers

Probers poke their beaks into large and small crevices to extract invertebrate morsels. The Brown Creeper of North America is a small bird with a sharp down-curved bill. Flying to the base of a tree, it works its way upward, spiraling around the trunk searching under and around the bark for insects, their larvae or eggs, and other small creatures. Its long claws and spiky tail feathers move the bird upward as it clings close to the tree appearing to be a piece of bark. Nuthatches operate similarly but with a sharp straight bill, a short tail, and a habit of starting at the top of the tree and working their way down. Although mainly insectivorous, nuthatches get their name from "hatching"—cracking—nuts and seeds open with their sharp bill. Like a few other birds, nuthatches will cache food, both insects and seeds, but typically with only one type of food item per cache.

Woodpeckers probe by sticking their bills deep into crevices, whether tree bark, roof shingles, or the ground. A most interesting arrangement allows the tongue to extend out perhaps four times the length of the beak. Some woodpecker tongues cover the hyoid apparatus, a series of muscle-covered bones in the shape of a Y with the horns of the Y originating forward of the nostrils or in the orbits of the eyes, looping over and down the skull, coming

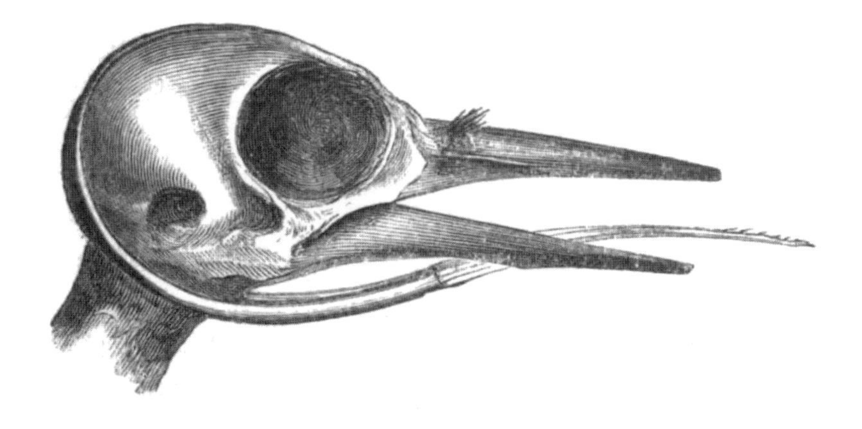

Woodpecker skull showing route of tongue.

together as one bone in the throat, and extending out between the jaws as the tongue. Depending on the species, the tongue may be sticky, have barbs or spines, or be flattened at the end.

Pecking and probing, woodpeckers also pound, and the perennial question is why they don't get headaches. Three reasons: the skull is a spongelike matrix of bones that helps to absorb blows; the lower jaw bends a bit, lessening the shock; and the hyoid bone acts like a seatbelt, preventing excess movement of the head. New helmets for Army tank drivers and bicyclists are being designed based upon the ingenuity of this woodpecker trait.

Shorebirds have made probing a real science. Probing into the mud for invertebrates such as worms, insect larva, amphipods, crustaceans, and mollusks, their bills go shallow, deep, or somewhere in between, divvying up the food sources by depth. Short-billed plovers probe down to a depth of 2 inches; medium-billed stilts feed at intermediate depths; and the long-billed curlews feed from the surface down to 8 inches. How does the curlew find and snatch its prey? Think about how difficult it would be to completely open a long pair of forceps in thick muck; long-billed shorebirds have figured this out—only the tips of their bill open and close—sort of like those picker-upper gadgets one uses to collect litter from the curb. Avocets, with

A variety of shorebird bills.
1. Spotted Sandpiper 2. Long-billed Curlew 3. Marbled Godwit
4. Stilt Sandpiper 5. Snowy Plover 6. Black-necked Stilt

long up-curved bills, sweep their bills side to side through the water or silt in search of prey or swim to deeper water and dabble, head down, like ducks.

But how do shorebirds find food in the first place, being that it might be buried in sand and muck? Recent research has shown that many shorebirds such as the Western and Least Sandpipers and the Red Knot can detect prey under the sand of the shoreline without seeing, smelling, or touching it. Herbst corpuscles, mechano-sensory organs just underneath the surface of the bill, are the answer. Sticking their bills less than a half-inch into the sand, the birds can detect the presence of a rock or a prey item because the corpuscles detect differences in the pressure gradient. We used to think that the nocturnal kiwi of New Zealand located its worm prey by smell since its nostrils are located near the tip of the bill and the olfactory lobes of the brain are large, but recent research indicates that kiwis have mechanoreceptors similar to that of shorebirds. Ibises also possess this adaptation and as the birds are not closely related, these mechanoreceptors must have evolved independently.

Oystercatchers live in almost all coastal areas of the world and feed on a wide variety of prey—on land they probe for worms and insects and in the shallows they poke about for marine worms and shelled invertebrates. But their long bills also serve as wedges to pry chitons and limpets off rocks or to crack their shells. Oystercatchers are named for their style of feeding on mussels and oysters by sticking their bills between the valves of a feeding mussel and cutting the posterior adductor muscle to open the shells. Young oystercatchers must learn this technique, often stealing food from adults before their education is complete.

Of all the probers, hummingbirds and sunbirds are probably the most recognized. Sticking their beak deeply into a flower, they extract nectar with their fringed and grooved tongues. There are 132 species of sunbirds and almost three times as many hummingbirds, so this lifestyle is obviously successful. Because birds and flowering plants came into being about the same time, during the Jurassic period about 150 million years ago, flowers and flower-probing birds have a close connection. Many flowers, actually their petals, have adapted a form and function to attract pollinators and reward them with nectar. Hummingbird-pollinated flowers tend to be yellow, orange, or red; have only a mild scent; are above average in nectar

Green Violetear Hummingbird supping nectar from a *Salvia* (sage) flower.

concentration and amount of pollen; hang horizontally or downward; and are tubular in shape. Hummingbirds hover and transfer pollen with their beaks; the *Heliconia* (false bird-of-paradise) flowers of Southern Hemisphere rainforests have sticky threads of pollen that adhere to hummingbird bills to make transfer even more efficient. Because hummingbirds feed while airborne, hummingbird-pollinated flowers can be located almost anywhere on the plant. But sunbirds sit while they feed, so sunbird flowers provide a perch. The rat's tail (*Babiana ringens*), a flowering plant endemic to South Africa, grows a spike specifically as a bird perch. Some sunbird-pollinated flower species bloom close to the ground to make their flower parts more accessible. And since they perch while feeding, sunbirds often transfer the pollen to another plant by transporting it with their feet.

FRUGIVORES: THE SEED DISPERSERS

Animal distribution of pollen is a much more efficient system than wind and exemplifies coevolution nicely. Pollination leads to fertilization and the development of seeds, which also need to be dispersed, so seeds are often encased in fruit to attract seed dispersers. Birds eat the fruits, and the hard-cased seeds exit the digestive tract or are regurgitated; the plant's seeds get moved away from the parent plant and the bird gets nutrition. Everybody wins. Fruit eating is common among birds; in one study in Costa Rica, 70 bird species were seen feeding on the fruit of 170 species of plants. Birds choose fruit based upon its nutritional value, seed-to-fruit size ratio, taste, time of ripening, and color. Berries are an important, sometimes sole, source of food in the winter when insects are scarce; the Mistle Thrush of England will defend holly berry patches and the Townsend's Solitaire of North America will defend stands of juniper trees. The Yellow-rumped Warbler is the only warbler able to digest waxy myrtle and bayberries, allowing the birds to survive the winter farther north than any other warbler species of North America.

Bird-dispersed fruit tends to be red but may also be blue or black, whereas mammal-dispersed fruit is generally yellow, orange, or brown. You would think that brightly colored red fruits would be most appealing to birds because they are to us, but experiments with Redwings in Finland indicate that blue fruits are attractive as well because birds see ultraviolet, UV light being reflected by some blue berries. In an area with few fruit-eating birds, fruits tend to be multicolored to enhance their attractiveness.

Birds that usually eat smaller fruits whole defecate the smaller seeds away from the tree, whereas birds that tend to eat the large seeds of large fruits at or near the tree drop them underneath rather than dispersing them. But since large seeds are more successful at germinating, and it takes fewer seeds to establish a new plant, the strategies are more or less equivalent. Most birds pass the seeds a few hours after ingestion, so the seeds are not deposited a great distance from the parent plant. However, two shorebirds—the Killdeer and Least Sandpiper—may retain seeds of bindweed, mallow, and sumac in their guts for 6–14 days. As long-distance migrants, these birds may deposit seeds thousands of miles from their origin.

Cedar Waxwings live mainly on fruit and are so efficient at digesting that they can pass mistletoe and other berries through their digestive tract in 16 minutes. They will gorge themselves, filling their throats and crops with berries until they move down the digestive tract. Sometimes the birds are so

Although Cedar Waxwings devour insects for protein during breeding season, the vast majority of their diet consists of berries.

satiated that they will pick a berry and pass it down a line of birds until one decides to eat it. The story is often told about waxwings gorging on fermented berries and getting drunk. It happens, not only to waxwings but blackbirds and robins as well; the ethanol in fermenting berries causes the inebriated birds to fall from trees or fly into objects, often causing their demise.

The fruits of some plants have evolved to be bigger, tastier, more nutritious, and generally more attractive to birds for dispersal, but other plants have toxic seeds to deter birds from digesting or damaging their seeds. *Nandina*, an ornamental plant commonly called heavenly bamboo, which is native to Asia but found in some parts of Europe and the United States, is poisonous to birds as parts of the plant, including the berries, contain cyanide. Parrots have enormously strong bills and can decimate and digest seeds, even those that contain toxins such as alkaloids (cacao), cyanide (apple), and persin (a fungicidal toxin in avocados.) Why aren't the birds affected? Parrots eat clay along South American river banks and one explanation is that the clay binds with the toxins in the digestive tract and renders the poison ineffective.

Coevolution is a dynamic process in all biological communities. A study in the Brazilian forest, dwindling in size because of logging, found that the size of the seeds of several species of palm trees has been shrinking. With the reduction in the number of large birds such as toucans because of habitat loss, palm trees are producing smaller seeds to attract smaller bird dispersers.

But smaller seeds could mean bad news for these tree species as the seeds contain less nutrition and are less tolerant of drought.

Only 3 percent of birds eat leaves (folivory) with any regularity. The leaf-eating Hoatzin of South America is an especially unusual bird. Its diet is primarily leaves, which it ferments, as in a cow stomach, in its crop. The crop is so large and well developed that the bones and muscles of the chest have been compromised, making the bird a poor flier. The young are fed with partly digested regurgitated leaves, and if that isn't unusual enough, the newly hatched birds possess two claws on each hand with which they can clamber through the vegetation.

CARNIVORES OR RAPTORS

Perhaps the most impressive of all birds are the large carnivorous ones—hawks and eagles, falcons and kestrels, hobbies and kites. From the Latin *rapere* for thief or plunderer, all these birds, which hunt and eat other animals, are called raptors. Raptors themselves have few enemies and tend to

The Hoatzin is the only member of the family Opisthocomidae—the family name is Greek for long hair in back, which refers to the bird's crest.

survive for many years, but like most birds, the first year of life poses the most challenges and fewer than half of their young survive to their first birthday. Raptors have numerous adaptations for a predatory life: strong legs and feet, sharp talons, and exceptional hearing and eyesight. Excellent vision is especially important when flying in pursuit of a meal. You would think a predatory bird would focus on and head straight for its prey. But observations of many hawks and falcons indicate that at some distance from their quarry, the birds fly with their head at a 45-degree angle while focusing on the prey. This requires that they fly in a spiral pattern to approach the victim; when less than 26 feet away, they straighten their head and fly directly at the quarry. Even though flying in a spiral takes longer and requires slower flight, Vance Tucker, a pioneer of bird flight studies, found that focusing on the prey at an angle insures a greater success rate than had the bird approached its meal straight on from a distance.

Hawks and eagles are essentially the same although the larger ones tend to be called eagles. There are about 60 species of eagles, mostly distributed in Eurasia and North America. They feed on live prey—mice, rabbits, skunks, gophers, groundhogs, snakes, other birds, and large insects. The Golden Eagle tends to hunt near the ground and swoops down upon its prey with talons forward and wings back, snatching a poor bunny off the ground and shredding it into edible pieces elsewhere with its sharp beak. The long claws and muscular legs of the Golden Eagle have allowed it at times to carry off foxes and coyotes and even lift a small deer off the ground. In Alaska eagles occasionally prey upon young Dall sheep and caribou and young domestic sheep from ranches. The Snake-eating and Serpent-eating Eagles of Africa prey mainly on snakes and don't avoid venomous ones even though the birds are not immune to the venom. The Rufous Crab-hawk of South America eats only crabs, using its long legs to snag them. The Bald Eagle is more of a fish catcher, but it might make five or six attempts before successfully nabbing one, sometimes taking a swim in the process. Bald Eagles will also eat mammals and birds and won't hesitate to steal a fish from the smaller Osprey. They will eat carrion and are often seen in garbage dumps in Alaska, where there are no vultures. Symbol of the United States since 1782 (winning the vote over Ben Franklin's choice of the turkey), the Bald Eagle, not bald at all, gets its name from the patch of white on its head and tail, a description known as piebald, meaning large spots or patches.

Hawks, from the Old High German *Habischt* (seizer), eat all kinds of animal prey. Searching for prey usually requires looking downward, and as we all know, bright sunlight can interfere with vision. So hawks have a bony ridge above their eye, located behind where an eyebrow would be, that shades their eyes like a built-in baseball cap—and also gives them an ominous look. Broad-winged soaring hawks are also known as buteos, after their genus name, and the accipiters (usually called sparrowhawks or goshawks) are the narrow-winged, fast-flying hawks. Buteos such as the Red-tailed Hawk of North America and the Common Buzzard of Europe ply the skies using updrafts as much as possible, flapping little, and looking for prey. The Red-shouldered Hawk leans toward lizards and snakes while the Red-tailed Hawk prefers skinless small mammals—the birds strip mice and gophers naked before swallowing them. The accipiters have long tails and zoom adroitly through the woods. Cooper's Hawks and their little brother look-alikes, the Sharp-shinned Hawks, are active fliers and hunt both from perches and on the wing. They are faster than the soaring hawks, feeding on both rodents and birds, but are most admired for their agility in pursuing songbirds through the brush. Autopsies of Cooper's Hawks indicate that a third of them had broken one or both clavicles at some time, most likely by flying into a branch in pursuit of avian prey.

Owls, nocturnal carnivores, share many similarities with hawks. They survive by catching and decimating small- to medium-sized prey with their muscular legs, strong feet and claws, tearing beaks, and good eyesight and hearing. Eating whole birds, mice, and gophers means the bird swallows hard-to-digest parts. About 10 hours after their meal, owls (as well as hawks, grebes, loons, cormorants, and others) regurgitate a round or oval pellet of indigestible items such as bones, feathers, fur, plant parts, and even bird bands from an unfortunate chickadee or sparrow that served as a raptor's lunch. When regurgitating the pellet, the owl will make the sound and movement of retching and wear an understandably pained expression. Owl pellets contain fur and feathers because owls don't pluck or skin their prey like hawks do—they just gulp the creature down. Owl pellet examination to

The Sharp-shinned Hawk is named for its sharply keeled lower legs.

determine an owl's diet is so common in schools and nature centers that you can buy pellets commercially, selecting for size and content (choose mole, bird, or pocket gopher). Barn Owl pellets are the best as their stomach juices are less acidic than other owls and so the pellets contain the finer bones.

Most owls have a varied diet, but some are specialists. The biggest owl in the world, Blakiston's Fish Owl of far eastern Asia, stands on the edge of a hole in the ice and waits for potential piscine prey. Pel's Fishing Owl of Africa is well equipped for dipping its talons into a river after prey as it can swing one of its three front toes to the back so that two sharp claws are positioned in front and back in order to grasp a fish. Most owls have extensions of feather barbs on their flight feathers to allow for silent flight, but fishing owls do not because they don't need to be quiet to hunt for fish. Pygmy owls and owlets are avid bird-eaters; Pygmy owls are known to take hummingbirds while the little guys are at feeders. Tawny, Long-eared, and Barn Owls account for 11 percent of the mortality of bats in the United Kingdom. Smaller owls of the tropics prey most on insects, especially moths, beetles, and crickets. Several kinds of owls eat snakes at times, but the Western Screech Owl of the southwestern United States occasionally brings a Texas Blind Snake to its nest, alive. The snake, which normally feeds on ant and termite larvae, will eat insect pests in the nest such as maggots (fly larvae). Young owls in nests with snakes grew faster and had a higher survival rate than snake-free nests.

Falcons, with narrow and tapered wings, are the fastest raptorial birds and include the kestrels, hobbies, and the well-known Peregrine Falcon, which is reputed to be able to exceed 200 mph in a dive. The Peregrine is found virtually everywhere in the world except Antarctica. Its diet is almost exclusively small birds. I once observed a Peregrine with a live bluebird in its talons; the falcon plucked the songbird totally naked, ate its head, and delivered the body to two young falcons in the nest. Because of their speed, agility, and beauty, Peregrine Falcons are popular with falconers. The species name *peregrinus* (Latin for wanderer or traveler) refers to the habit of falconers in the Middle Ages catching adolescent birds en route to their breeding grounds. The genus *Falco* (from the Latin *falx)* refers to the sickle-shaped wings (or talons or beak).

PISCIVORES: DIVERS, SKIMMERS, AND MORE

Piscivores, fish eaters, include loons, grebes, penguins, murres, guillemots, puffins, terns, cormorants, anhingas, gannets, and Ospreys. The Northern Gannet is a world-class fish catcher, diving straight down into the sea as deep as 75 feet. Well-adapted for diving, they have no external nostrils, a strong sternum, and special air sacs, not unlike bubble wrap, that act as air bags to cushion the shock of hitting the water. Unlike the diving gannet, the Osprey (from the French *ossifrage*, bone breaker) can catch fish only near the water's surface. A bird of worldwide distribution, the Osprey catches fish like Pel's Fishing Owl by swiveling a toe around from a three in front and one in back perch position to a two and two hunting arrangement. Hunting on the wing, the bird flies at different heights to locate prey and then dives rapidly with its wings held back and talons forward. Ospreys are successful about one-third of the time, and their sharply curved talons and sharp spines on the bottom of the toes assure that the slippery prey does not get away.

Especially odd birds are the skimmers, a small family that includes only three species. The Black Skimmer opens its mouth, lowers its longer lower mandible into the water, and skims the surface for fish and squid along ocean shores. When it detects an edible item, it rapidly and reflexively grabs the prey in its bill and tosses it up for swallowing. A fast-growing rhamphotheca compensates for the wear and tear on the bill. With this seemingly energy-intensive feeding strategy, their day-to-day survival is assisted by their tendency to fly at a moderate speed, taking advantage of the uplifting aerodynamics of the water's surface, and feeding at night when more fish and squid come close to the surface.

At least as strange is the anhinga, sometimes called snakebird because it often appears with only its head and skinny neck above water. Although the anhinga has a functional uropygial gland—the oil gland on the rump of birds that produces the substance used to waterproof feathers—the anhinga's feathers are three times as wettable as other aquatic birds. They are able to

The Black Skimmer at work.

swim underwater easily and, with their very flexible neck, can spear fish, toss them upward, and swallow them intact.

FILTER FEEDERS

Flamingoes remind me of the scene in Alice in Wonderland when the Queen of Hearts uses a flamingo as a croquet mallet. These unusual birds utilize rows of horny plates in their jaws to filter out microorganisms from water—sometimes so filthy looking that it's hard to imagine it supports any life. Flamingos feed with their head upside down, the lower beak pushing up against the upper. The thick tongue serves as a pump, moving back and forth as frequently as four times a second, pushing microorganisms against the filter mechanism while the bird sweeps its head through the water, the filtered water gushing out the corners of the bird's mouth. The pink color of the leg, face, and feathers comes from their diet of carotenoids—pigments in crustaceans and some microorganisms. The carotenoids are deposited in the integument (such as skin or feathers) of the bird and when they dissolve in fat molecules, the reddish color is exposed. This is the same thing that happens when you boil a lobster.

Dabbling ducks, those that tip down and stick their butt up while floating, use the lamellae (ridges) on their bill edges to filter out seeds and small organisms from the bottom of a lake or pond. The Northern Shoveler has especially well developed lamellae that extend into fine hairs for exceptional filter feeding ability. Unlike other dabblers such as the Mallard and American Widgeon, the Northern Shoveler can also sieve organisms from the water's surface. The spoon-shaped bill of the shoveler has earned the bird its colloquial names of Hollywood Mallard, Smiling Mallard, Daffy Duck, Daisy Duck, and Spoonie.

VULTURES AND RELATIVES: THE FULL-TIME SCAVENGERS

Many birds are occasional scavengers like the Bald Eagle, gulls, pigeons, and members of the crow family. But the birds that eat dead creatures as a living are the most fascinating, although most birdwatchers would probably not count them as their favorites. Many people think that vultures circling overhead means that something is dead below them. But in the event of a deceased animal on the ground, the vultures would be there, feasting, not soaring above and drooling. The birds are circling because the wind conditions are such that it's easy to stay aloft and search for corpses without too much wing work. Impress your friends the next time you see vultures circling by calling the formation by its proper term "kettle," as if the birds were being stirred.

The New World vultures are in the family Cathartidae, from the Greek *skathairein*, to purge or cleanse. The word could refer to the fact that vultures rid the environment of dead animals, or to their habit of regurgitating stomach contents upon the approach of a predator to lighten the load for takeoff. In any case, they eat dead things, which are undoubtedly full of bacteria. To counter potential infection vultures have strong stomach acids to kill pathogens and they defecate on their feet both as an antiseptic wash and, on hot days, to help cool them off. They have naked heads as a feathered head would be impossible to clean; a naked head also allows sunlight to disinfect the skin. Most vultures of the New World have an excellent sense of smell and detect ethyl mercaptan, a gas produced by decaying bodies. That's a handy survival mechanism, but what about Black Vultures whose sense of smell is not nearly as well developed? They fly above Turkey Vultures and follow them to a meal of carrion; they will also prey on young or injured mammals.

Although they generally look and act alike, New World vultures (from the Americas) are not closely related to the vultures of the Old World (Europe, Africa, Asia). One big difference is that Old World vultures do not have a good sense of smell. Since they need to find their food by sight, African vultures fly considerable distances each day, often following ungulate herds. More than 80 percent of the time they find food by joining a group of vultures that has already begun feeding on a large carcass. As they converge upon a carcass, which I'm sure you have seen in nature films, they will squabble, but these groups have developed ritualized displays to establish a dominance hierarchy to minimize aggression. Many of the carcasses of large ungulates that vultures feed on died of disease, starvation, or both; although vultures might spread some bacteria via their feathers, they do a pretty thorough job of ridding the environment of diseased carcasses.

Vultures and condors are at risk worldwide because of the contamination of their food sources. Lead shot left by hunters and poisons used to kill predators or pests are often found in the carcasses of animals that the vultures consume. A spectacularly sad case is that of the White-rumped Vulture in India whose population numbered in the tens of millions in the 1980s; today only a few thousand survive. The culprit is Diclofenac, an anti-inflammatory drug given to cattle that is poisonous to vultures. Cattle are mainly used for milk and rarely eaten in India, so millions of carcasses

The Griffon Vulture is an Old World vulture found across the Mediterranean into the Middle East and Asia.

are available to vultures, India's primary animal disposal system. Even with their tough and resilient lifestyle, vultures cannot survive this drug that shuts down the kidneys.

When you see a bird in the wild, it is usually searching for, manipulating, eating, or carrying some sort of food item. Although birds may not always be initially successful in establishing or defending a territory, finding a mate, building a nest, or raising young, they will often get another chance. But if they are not successful in foraging, they have no chance at reproduction and their genes will be lost from the gene pool. The everyday habits of birds, circumscribed by the shape of their bill and their foraging behavior, provide essential information in the study of avian species. Watch your bird feeders and marvel at how evolution has shaped what you observe, because the beak and feeding habits you see garner the fuel for all the other activities birds need to survive.

CAN YOU SEE UV?

How Birds Employ Their Sensory Abilities

In those birds whose eyes are placed "laterally," that is, on the side of the head, the two eyes are used for different tasks. Day-old chicks of the domestic fowl, for example, tend to use their right eye for close-up activities like feeding and the left eye for more distant activities such as scanning for predators.

—TIM BIRKHEAD, *Bird Sense: What It's Like to Be a Bird*

We are constantly processing input from the environment, primarily visual and auditory signals, but we tend not to be acutely aware of our surroundings unless the stimuli are fast, colorful, or loud—in situations like crossing a busy street or playing tennis. Birds, on the other hand, appear to be exquisitely sensitive to their environs at all times. Have you ever approached friends who accused you of sneaking up on them? It has been a rare experience for me to successfully approach a wild bird as they seem to be much more aware of their surroundings. We humans miss or misinterpret signals with few repercussions. For birds it could mean life or death.

A bird's sensory organs work basically the same as ours, but with different qualities and sensitivities to help them do things like find amphipods under the sand, see UV, or navigate over the ocean without instruments. The more we learn about birds, the more we uncover their extraordinary sensory abilities.

SENSE AND INTELLIGENCE

"Bird brain," that negative appellation comparing someone's lack of intellectual abilities to that of a bird, totally misses the mark. When I entered the field of ornithology as a grad student, the thinking among us zoologists was that mammals were smarter than birds because they have a relatively bigger cerebral cortex (controlling perception and behavior) while the cerebellum (regulating fine motor movements) is bigger in birds. The logical conclusion was that mammals are good learners but clumsy and birds are instinctive but agile. We were mostly wrong. It is safe to say that the cognitive abilities of birds, their social behavior, and their exceptional skill at discriminating among visual images make them a lot more capable than we thought possible, especially groups like the crows and parrots.

Birds receive input from their senses and then "decide" what actions to take. Ornithologists, behaviorists, and neuroscientists are still trying to tease out how nature and nurture interact, but it's clear that both learned behavior (such as choosing the right food) and innate actions (like escape), often modified by learning, are based on information coming into the brain of a bird via its various senses. Young birds have to learn what to fear. Intuitively snatching every insect culminates in a learning experience when a toxic bug shocks the taste buds; blowing leaves frighten fledgling birds, which instinctively fly off as if the leaves were predators.

Tool-using birds are represented in 33 families. Some crows, warblers, finches, and woodpeckers fashion twigs to pry insect larvae out of tree bark; Egyptian vultures use rocks to break open Ostrich eggs; and crows drop walnuts in traffic for cars to crack. In the early 1900s, Blue Tits in England took to drinking cream off the top of home-delivered milk. When the milk producers covered the bottles with aluminum foil, the birds learned to poke holes through it. This behavior has an innate component, but through experience the birds learned to refine their techniques. Birds have survived for more than 150 million years by incorporating new abilities into their genes as the environment changed.

We need to be cautious in assuming that we know how birds see the world. The actions of our own friends can sometimes be puzzling—and we know what senses they use to receive input from the outside world—so you can appreciate how much more difficult it is to understand the world from a bird's perspective. In Manaus, Brazil, where the Rio Negro flows into the Amazon at the "meeting of the waters," thousands of Purple Martins roost at night during the Northern Hemisphere winter in a large oil refinery, just across the river from the rainforest. Steam, lights, and noise from the pipes mix with the cacophony of bird calls. Do the lights and noise bother them? Do they stay there because they are safe from predators? To figure out why birds do what they do, we first have to discover how they perceive their environment.

VISION: SEEING IS SURVIVING

"Eagle-eye" has come to define someone with particularly good vision. Overall, birds have excellent eyesight, and no wonder—taking off, capturing prey,

The Purple Martin may benefit from roosting among the hot pipes of a refinery because the heat destroys the insect midges which carry parasitic protozoans that infect birds.

avoiding objects and predators, landing, and just hopping around require sharp vision. Vision is essential to birds not only because they are exceptionally active and mobile, but also because much of their communication is based on visible signals from other birds: distinguishing individuals of their own or other species, responding to courtship displays, recognizing territorial boundaries, and interpreting warning signals such as raised crests and alarm calls. Many animals do well in spite of their moderate visual capacity, making up for it with other abilities, but birds are successful largely because of their particularly acute vision and have exploited visually demanding habitats more effectively than any other animal group.

Eye Size, Shape, and Color

With eyesight being the predominant survival sense in birds, it is no wonder that evolution has selected for large eyes, especially in larger birds. In hawks, owls, and other non-songbirds, the size of the eye appears to be constrained only by the size of the brain with which it shares space in the skull. For owls, the eyes might constitute 1 to 5 percent of the bird's total weight. Some observers have suggested that the face shape of owls is actually a result of their large eyes, which nearly touch in the midline. Ostriches have the largest eyes of all land animals, measuring 2 inches across and weighing nearly as much as the bird's brain. The smaller size of songbird eyes is related to the bird's body size rather than brain size, but their eyes are still large compared to ours. The European Starling's eyes comprise 15 percent of the weight of the head, whereas the weight of our eyes is only 1 percent of our skull and contents. The larger the eye, the more light is gathered, so nocturnal foragers, like some shorebirds, owls, and nighthawks, have larger eyes. Fast-flying birds like falcons, swifts, and swallows—which need acute eyesight to avoid obstacles—also have larger eyes.

Birds that don't need particularly sharp vision like sparrows, quail, and pigeons have "flat" eyes, which means the distance across the front of the eye is greater than the distance from front to back. This design compromise both captures the maximum amount of light and provides for wide-angle vision, but the small image that falls on the retina results in low visual acuity. Hawks, most eagles, falcons, and some songbirds have rather round eyes that narrow the field of vision, but allow more light to fall on the retina, increasing visual acuity. Owls and some eagles have tubular eyes. This shape allows the birds to focus intently straightforward, and because the birds have little or no peripheral vision, they tend to fly in a level flight path. Partly because of this habit, vehicles too often hit owls as they fly across a road. Some studies indicate that roadkill is the greatest source of mortality of Barn Owls in the United States.

The lenses of bird eyes are more variable in shape and accommodation ability (the change of lens shape to allow for better focus) than those of any other vertebrate group. A series of thin, overlapping, bony plates called the sclerotic (from the Greek *skleros*, hard) ring surrounds the eyes of all birds. When the strong ciliary muscles change the shape of the lens while focusing, the sclerotic

The Barn Owl's eyes are twice as light sensitive as a human's.

ring maintains the shape of the eye. The sclerotic rings of owls are almost like cylinders encasing the eye, maintaining a tubular shape and precluding eye movement.

Iris colors are also extremely varied, much more so than those of humans, probably because eye colors are often clues to a bird's species, age, or sex. For example, female Brewer's Blackbirds have black eyes while the males have yellow eyes. Because the female's plumage is much duller than the male's they can easily distinguish sexes so the yellow eye of the male may simply be an attractive feature to the female. The eye color of the Eastern Towhee is white and the Spotted Towhee is red, perhaps an important distinguishing sign where their ranges overlap. Sharp-shinned and Cooper's Hawks nestlings have gray eyes, young adults yellow, mature adults orange, and older adults red; eye color stages may be involved in maintaining a hierarchical structure.

Field of View: A Bird's Observable World

Humans, looking ahead, are always moving forward into their world (often looking down into some handheld electronic gizmo), but birds, with a substantial field of view, are surrounded by their world and with frequent head movements, take it all in.

Human eyes have about a 160-degree field of view, the extent of the observable world at any moment. With an eye on each side of the head, songbirds have a nearly complete field of vision (except directly behind). Directly in front, they have 20–30 degrees of binocular vision, providing depth perception, but most of their sideward vision is monocular, which precludes depth perception but is sufficient to detect the approach of a predator. Ducks like the Mallard can see nearly 360 degrees, with a narrow binocular view in front and a blind spot in the back of their head. The American Woodcock can see 360 degrees, including narrow slits of binocular vision in the front and back of its head since the woodcock spends a good deal of time facing down and probing the ground.

Because birds have limited eye movement, they scan their field of view with neck movements. That's why you see birds bobbing their heads up and down, looking quickly side to side, and craning their necks. We can turn our heads another 160 degrees to scan our surroundings, but any more than that and we would cut off blood flow to the brain as muscles and vertebrae would constrict blood vessels. Owls don't have this problem. The field of view of an owl facing straight ahead is about 70 degrees with binocular vision and about 30 degrees additional on each side with monocular vision. But they have large holes in their neck vertebrae that provide extra space for the arteries traversing through them, allowing the birds to turn their head a total of 270 degrees to scan their surroundings.

Flexible necks make up for fairly rigid eye positions, but birds still have some eye movement. Pigeons, like most birds (at least that's the assumption because there haven't been enough experiments), move their eyes simultaneously in the same direction. That is, if one eye moves toward the front, the other does as well. Makes sense, yes? But experiments with Zebra Finches show the contrary. When one eye moves forward, toward the beak, the other moves backward the same distance, and vice versa. So how does the brain cope with two very different images? Turns out that whatever stimulus (food, another finch) appears on one side or the other, that eye moves toward the stimulus and the other eye responds by moving in the opposite direction. Presumably, this allows the Zebra Finch to attend to the object of attraction while still getting input about predators, for example, from the other eye.

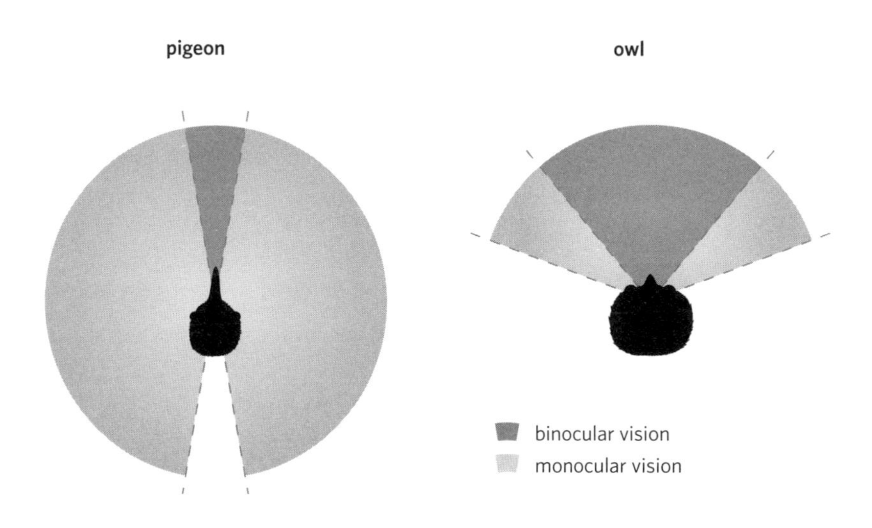

The fields of view of a pigeon and an owl.

One would think that a wide binocular field of view would be essential for flight—and the fact that the flightless kiwi has only a 10-degree binocular field would seem to support that—but in the majority of birds only 15–30 percent of their field of view is binocular. However, woodcocks that fly quickly through dense wood and filter-feeding ducks that zip through wetlands and nest in heavy vegetation have a comparably small field of binocular vision to that of kiwis. The birds make up for their narrow binocular vision and avoid obstacles by being able to compare the images from each eye. If one eye sees a tree branch move faster than the other eye sees it, for example, the bird senses it is moving toward the branch and adjusts its flight in the other direction.

Try this: put your hand in front of your face with your arm extended. You can see your hand clearly. Now move your hand slowly to the side but keep looking straight ahead. Eventually your view of it becomes less clear and ultimately unrecognizable. In the center of your retina is the fovea (Latin for pit), which contains a high concentration of sensory cells and is responsible for acute eyesight. As your hand moves to the side, less light from it hits the fovea so the image of your hand blurs. The fovea is highly developed in birds, and about 50 percent of species, especially those that depend on sharp vision, have two foveas: one fovea looks straight ahead, like ours, and another looks to the side. So a bird can look to the front and side at the same time, important while pursuing prey. Most shorebirds have one ribbonlike fovea that focuses in a strip, appropriate for the flat horizon they see. Nocturnal and crepuscular birds, like owls and frogmouths, have only one fovea.

Flicker Fusion

When birds are flying, objects whiz by their fields of view. Recall the old silent films during the Charlie Chaplin era in which everyone seemed to walk in a jerky fashion? That's because those films were shot at 16 frames per second and our vision perceives a bit of a gap between each individual frame. Today films are shot at 24 frames per second, and TV cameras project at about 30 times per second. At those speeds we do not notice the change from one frame to the next, making the action appear seamless. This is called the flicker fusion rate. When driving we don't see objects quite in focus because our brain didn't evolve to discern images so quickly. But it did for birds. The studies done on birds—mostly pigeons and chickens—indicate that their flicker fusion rates are more than double that of humans. That means if birds were to watch movies, they would have to be shown at perhaps 100–120 frames per second or the birds would see the flickering of individual frames. Another example: humans generally do not notice the flickering of fluorescent lights, but researchers working with chickens indoors found that the birds can distinguish individual flickers of fluorescent lights. Although humans might develop symptoms such as headaches, eyestrain, nausea when exposed to flickering lights, the chickens are apparently not bothered.

When we see a flock of birds turning rapidly in midair, we view it as a coordinated, almost balletic swarm. But if we were to slow down a high-speed film of an airborne flock it would look much more chaotic, like a frantic mob rather than a flock. Because avian eyesight is optimized for speed, birds can more readily distinguish fast-moving objects and individual birds. Thus they are able to keep track of their adjacent flock mates and usually avoid collisions.

Eye Protection

Flying requires good vision but it also subjects the eyes to fast-moving air. To protect their eyes, birds have a nictitating membrane or nictitans (from the Latin *nictare*, to wink or blink), in addition to eyelids. This "third eyelid" performs the function of rapidly cleansing and moistening the eye. In many birds and reptiles the membrane is lined with feather epithelia, club-like extensions of skin cells similar to feathers that help clean and refresh the corneal surface. The eyes of flying birds, especially those of fast-flying falcons, are subjected to drying air, so surface lubrication via the nictitans is important. Woodpeckers, adept at hammering away at tree trunks, face special physical problems, a couple of which are solved by the nictitans. First, the membrane acts as a safety belt, helping to prevent the eyes from bulging too far out of their sockets as the bird exerts 1000 Gs of force on a tree. Second, pounding on a tree produces woodchips and dust, but the membranes protect the eyes from any damage. The nictitating membrane of diving birds like loons and grebes protects their eyes from the drying effects of saltwater.

Seeing Color and Ultraviolet

To say birds see color seems pretty obvious because so many birds are so delightfully tinted with subtle shades of blue or red, highlighted with contrasting colors like black and yellow, decorated with a spectacular mosaic of rainbow colors, or bejeweled with iridescence. The common names of hummingbirds—Audubon called them "glittering garments of the rainbow"—include such accurate descriptions as azure, sapphire, emerald, brilliant, and sparkling. At the Currumbin Wildlife Sanctuary in Queensland, hundreds of green-backed, orange-chested, blue-headed, yellow-accented Rainbow Lorikeets greet visitors twice a day like the shattering of a colorful mosaic. And while sitting in a leaky boat with a rotten transom in the Caroni Swamp in Trinidad, West Indies, I caught sight of a tree full of Scarlet Ibises, the bright birds declaring their personal space on various branches, like ornaments on a Christmas tree. These hummingbirds, lorikeets, and ibises were certainly not selected by evolution for my entertainment, so why such colors?

Surprising as it first seems, color appears to be the driving force in early feather evolution. As discussed in the next chapter, bird feathers arose not as a modification of scales as was long presumed, but as novel structures on the dinosaur precursors of birds. Further investigation and fossil discoveries determined that even the most primitive forms of feathers were colored and had to be there for decoration. Today the purpose of colorful plumage in species, age, and sex recognition, courtship, and establishing and maintaining territories is evident.

Birds can see colors because they have rods and cones, sensory cells in the retina that transmit information to the optic nerve and brain. Rods detect black and white and are most numerous in nocturnal birds, whereas cones perceive color. Humans have three cones, detecting red, green, and blue. Birds have these cones, too, but they see more shades of color because their range is more evenly distributed across the spectrum. Birds also possess oil droplets in their cone cells that enhance the perception of color and sharpen vision by reducing glare, like polarized sunglasses. So our usual assessment of plumage color—red cardinals and pink flamingoes—is fine for describing birds, but may not reflect what birds see.

Along with red, green, and blue, birds have a fourth color receptor capable of perceiving ultraviolet (UV, "above violet")—the spectrum of light radiation that has a shorter wavelength but higher frequency and energy than visible light. One study on monomorphic birds (those whose sex we cannot distinguish) used instruments that measured UV reflectance and indicated that more than 90 percent of birds whose sexes appear identical to us are actually sexually dichromatic to the birds themselves. Another interesting study investigated the response of territory-holding Yellow-breasted Chats to an artificial model. Even though the sexes of the chat look identical to us, the territory-holding birds responded with sex-appropriate behaviors to the models, depending on whether they were made of female or male feathers. The only clue they could have responded to was some aspect of plumage colors such as UV reflectance.

The ability to perceive ultraviolet means that some birds can not only identify individual birds, but can find food items based upon varying amounts of reflected ultraviolet light. A variety of fruits contain anthocyanin, a pigment and antioxidant. As the fruits ripen, anthocyanin levels, the caloric value of the fruits, and UV reflectance all increase. The plants evolved

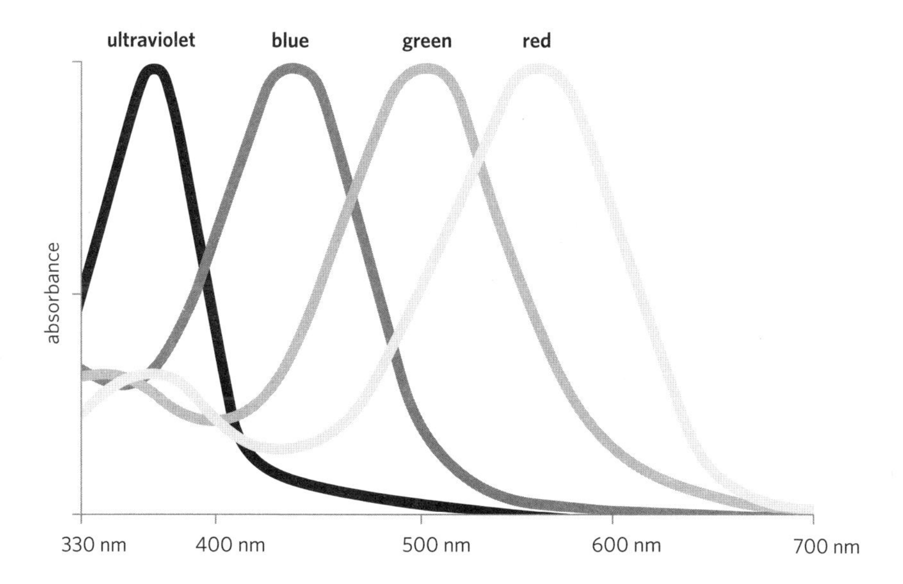

The range of sensitivity for each type of color receptor in birds—UV, blue, green, and red.

to be bird-dispersed, so ripe berries signal the birds with UV. Some predatory birds find their prey the same way. European voles are small rodents that mark their trails with urine as they move through the grass. Fresh urine has a higher UV reflectance than older urine markings so European Kestrels and Rough-legged Buzzards find their vole prey by following the fresher scent trails. A study of 98 songbird species in Europe found that the eggs of hole-nesting birds had more UV reflectance than those of open nests, presumably to help the parents see the eggs.

UV reflectance can also help European Rollers with large broods determine which of their young is most in need of nourishment. A relative of the kingfishers, European Rollers begin incubation soon after the first egg is laid, resulting in asynchronous hatching and thus different-sized hatchlings in the nest. The larger young tend to compete more successfully for food brought by the parents so the smaller young may receive less food. In a small brood, one to three young, there is little difference in the UV reflectance of the young, as the parents have no problem in provisioning for the whole family. But in a larger brood, thinner nestlings will reflect more UV light than the skin of larger, healthier nestlings, allowing the parents to fine-tune their feeding strategies to raise the most young.

Seeing Near and Far

Birds that forage in watery environments face interesting challenges. Because air and water have different refractive values (they bend light differently), animals that have normal vision in air become myopic (near-sighted) in water and animals that live in water are far-sighted on land. Gannets are one of the few birds that dive from considerable heights and then wing-flap underwater in pursuit of fish. Gannets are usually far-sighted when flying over the ocean, but within 0.1 seconds of entering the water, they become near-sighted by changing the shape of their lens to facilitate chasing their piscine prey. Kingfishers have two foveas in each eye. They use the fovea near the bill to focus on its aquatic prey item while diving toward it from a perch; upon entering the water, the more centrally located fovea comes into play so that the bird now has binocular vision. The Green Heron fishes in shallow water by dropping bait such as a twig, insect, or flower onto the water's surface and waits for its prey to rise and take a bite. Due to refraction, the fish appears to be somewhere that it really isn't, so the bird learns by trial and error to account for the redirected light from the prey. The oil in the color-perceiving cone cells also minimizes glare. The Tricolored Heron makes its life easier by shading the water and eliminating the sun's reflection as it dances through the water with wings spread as it follows and stirs up fishy prey.

Although penguins tend to be near-sighted in both air and water, their eyes appear to function equally well in both environments. They don't need to see very far on land and the suspension of particles in the water disturbs the clarity so that they cannot see very far while swimming. Underwater,

Tricolored Heron shading the water to attract aquatic prey.

penguins see green, blue, and violet best, but not red, probably because the low energy of red light does not allow it to penetrate deeply. The birds can probably see ultraviolet but it may be that the underwater vision of penguins depends more upon contrast than on color.

CAN YOU HEAR ME NOW? HOW BIRDS DEPEND ON SOUND

Hearing is better developed in birds than in any other terrestrial vertebrates. Diurnal birds use sound in many ways—to listen for prey, predators, warning calls, mates, competitors, calls from their young, and so on, and in combination with their eyesight, they are acutely aware of their environment. Nocturnal birds, however, depend much more on sound even though they have excellent night vision.

Like that of a mammal, the ear of a bird has external, middle, and inner sections, and an eardrum to transmit acoustical vibrations. Instead of the three vibration-transmitting bones of mammals—the malleus, incus, and stapes (hammer, anvil, and stirrup)—birds have only the columella (Latin for small column, which it resembles). When sound waves hit the ear, vibrations are transmitted from the eardrum to the columella to the cochlea, the fluid-filled, slightly curved organ that contains hair cells with their nerve endings. The hair cells move in the vibrating cochlear fluid, amplifying weak sounds and converting them to electrical signals to the brain via the auditory nerve. Unlike humans, whose hearing declines with age and exposure to loud noises because of damage to and loss of hair cells, birds maintain their hearing all their lives because avian hair cells continually regenerate—important to such aural animals. Since this discovery in birds in the late 1980s, research to regrow ear hair cells in humans has restarted.

The external ear of birds lacks a pinna, the cartilaginous flesh-covered outer ear like we have. We cannot see the ear opening of birds except for those with unfeathered heads such as vultures and storks. If you were to hold a parakeet in your hand, though, and blow gently on the side of its skull, you can expose the opening that leads to the inner workings of the ear. This hole, the external auditory meatus (Latin for passage), is covered by auricular feathers that protect it from rushing air as the bird flies and helps to funnel sounds into the ears. The auricular feathers of diving birds protect their ears from water pressure. Because the Ostrich and its relatives are not flying birds and ear protection is less important, they only have a thin covering of feathers over their ear holes. The structures of owls that are sometimes referred to as ears are actually feather tufts that play no role in hearing but instead indicate the bird's mood; the real ears are not visible.

California Condor; note ear opening behind eye.

Most birds receive sounds more or less equally because their ears aren't far enough apart for the separation of the sounds to be significant. Barn Owls have a flattish facial disk that funnels sounds toward the ears. The majority of birds determine the source of a sound by moving their head, like we do, and some evidence indicates that the ear closest to the sound registers the sound at a louder and higher frequency, helping to localize it. It has been said that if you see an American Robin or European Blackbird walking with its head turned toward the soil they are listening for their prey—worms or insects crawling in their burrows or under the litter. But in reality the birds are simply turning one eye to the ground to look for worm castings or other visual signs of prey.

Owls have fleshy, cartilaginous ears, not unlike ours, but asymmetrical in shape and location—they don't look exactly alike, and one is higher on the head than the other. Sometimes, when I give a talk on birds I ask for a volunteer in the audience to help explain why owls can locate sounds better than we can. I ask the volunteer to close her eyes and tell her I will snap my fingers in front of, over, or behind her head and that it is her job to determine the direction of the sound. Since I snap my fingers in the vertical plane that bisects her head from front to back, the volunteer rarely guesses correctly because the sound hits both ears with the same frequency and volume.

Volume and Frequency

Volume is the loudness or strength of a sound. Human conversation, for example, is rated at about 60 decibels, a dishwasher 80 decibels, and a motorcycle 25 feet away about 90 decibels. Whispering and the rustling of leaves only register at 20 decibels. A person with excellent hearing might be able to hear a sound at -15 decibels; Tawny and Long-eared European owls can detect sounds as soft as -95 decibels. (Seems curious, but decibels, like degrees of temperature, can be registered as negative.)

The sound-producing organ of birds is called the syrinx (Greek for pan pipes). Air moving over the syrinx produces a vibrating sound wave. The number of vibrations per unit time is the frequency and what our brains interpret as pitch. The measurement for sound frequencies is Hertz or Hz, defined as one cycle or vibration per second. Humans hear between 20 Hz (the lowest pedal on a pipe organ) to 20,000 Hz (a dog whistle), but are most sensitive to those between 1000 and 4000 Hz. Birds are most sensitive to sounds ranging from 1000 to 4000 Hz but there is considerable variability. In general, humans have a wider range of hearing than do most birds. The range of frequencies that a bird can most easily detect differs among bird species, and what birds can hear is closely related to the frequencies that species of bird can produce. The Horned Lark hears best between 350 and 7600 Hz, the Canary from 1100 to 10,000, the House Sparrow from 675 to 11,500 Hz, and the Long-eared Owl from 100 to 18,000 Hz, one of the broadest hearing ranges of all birds. Birds are more sensitive to frequencies, though: humans hear sound in bytes of about 0.05 of a second; birds hear the same sound in bytes of 0.005 of a second, so birds might hear 10 sounds in the time a human hears one. This allows birds to easily pick out sounds even in a rather noisy environment. The Hairy Woodpecker and probably other woodpeckers have the ability to hear beetle or bee larvae crawling under the bark of a tree. The Great Gray Owl can hear the rustling of a mouse below 13 inches of snow.

During my lectures on songs and calls in my ornithology class I analyzed a recorded bird song with a sonograph and showed the graphed printout of the sound frequencies the bird uttered. Then I asked students to try to imitate the bird sound and analyzed the best imitation with the sonograph. Although a student might produce a good imitation as judged by our ears, the graphs of the two songs were typically very different. Our vocal cords are not nearly as complex as the syrinx of birds and our hearing is not nearly as discriminatory. The sonograph of a Willow Warbler shows that the highest frequency the bird sings is at the beginning of the song and that it declines somewhat over a period of 4–5 seconds with a note sung about every 0.2 of a second.

Songs, Calls, and Why Birds Sing

Usually complex and relatively long, songs are those bird sounds that are considered melodious and pleasant to the human ear. Songs are most highly developed in the songbird group (order Passeriformes, also called the perching birds), which contains about 56 percent of all bird species. But not all members of the songbird group are singers; crows, jays, and jackdaws are

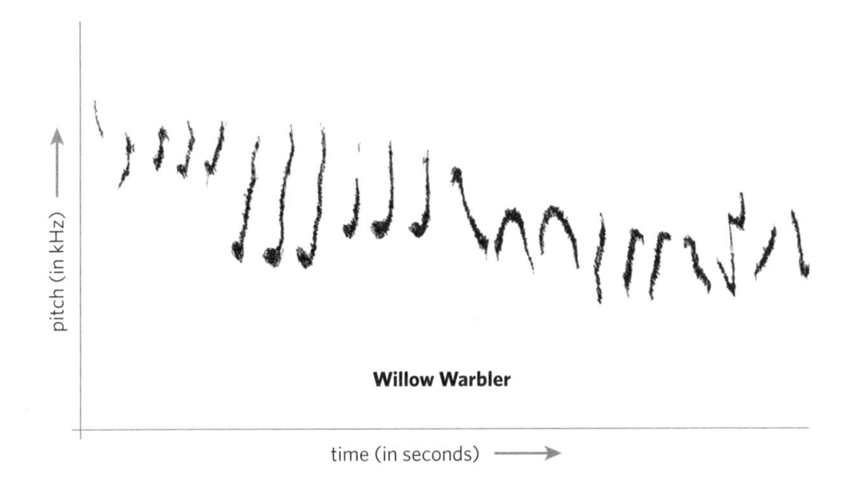

Willow Warbler song, indicating a slight decline in pitch over its duration.

certainly not known for pleasant warbling. Some naturalists construe bird song anthropomorphically, stating that birds sing because they are happy and enjoy singing. I always hate to disabuse them of this interpretation, but as a scientist, I know natural selection rarely allows frivolous behavior. The problem with the idea of a bird singing for pleasure is threefold. First, it's usually only the male singing; why not females? Second, singing is usually restricted to breeding season when courtship and mating are taking place. Does that mean birds are not happy the rest of the year? Third, singing advertises a bird's presence, which is ok if a female or competitor notices, but risky if a predator does. So, in the non-breeding season, a singer would just be announcing to predators he's there for the taking. Exceptions include the European and American Robins, which hold winter feeding territories and sing to defend them. Female Northern Mockingbirds, Northern Cardinals, and Black-headed Grosbeaks have songs as complex as the males, but again, typically sing only in the spring.

Calls are typically short, only a note or two or three, and are heard most of the year. Birds may use calls to communicate a threat, keep a flock together, indicate a food source, intimidate a predator, or announce their location. The number of distinct bird calls or songs depends on the range of the bird, the habitat, and neighboring birds, which might have a similar voice. The

Savannah Sparrow singing.

Common Raven has at least 25 calls. Chaffinches have a song plus 10 different calls: the flight call, social call, aggressive call, injury call, three courtship calls, and three alarm calls. Chaffinches mob a predator with low-pitched sounds described as "chink" calls. But when hidden in cover, the birds give a high-pitched "seeet" call, a thin note that causes other Chaffinches to seek cover. The chink note is easy to locate because of its low frequencies that differ in phase (sound vibrations of the same frequency occurring at slightly different times), whereas the seeet call is impossible to locate because it is composed of high frequencies with little phase difference.

Many birds have species-specific warning calls. Crows give a warning call that will frighten away only other crows. Nestling Dunnocks and European Robins only respond to the alarm calls of their parents by ceasing begging and hunkering down in the nest; reacting to the alarm calls of other birds would interfere with their feeding. But birds may also respond to the warning call of other species, whether or not the call resembles their own. A brochure from the Minnesota Department of Natural Resources states in bold print that "gull distress calls are specific to species and region." Nope. Great Black-backed and Laughing Gulls will respond to the alarm calls of

Herring Gulls. Fledglings just out of the nest suffer from a high risk of predation and immediately face the task of determining which alarm calls to respond to. Young Australian White-browed Scrubwrens, immediately after fledging, respond only to the alarm calls of adult scrubwrens. However, two weeks later, they also respond to the similar-sounding alarm calls of Superb Fairy Wrens and the very different alarm calls of the New Holland Honeyeater. But scrubwren fledglings in areas without honeyeaters do not respond to recorded honeyeater alarm calls. This demonstrates that the fledglings learn appropriate responses to the calls of other species.

Communication between parents and young begins even before hatching. Young quail, still in the egg, chirp to their mother and ask her to turn the eggs to the correct position for hatching if they are upside down. The chicks also chirp to each other; the older ones closer to hatching chirp slowly and the younger ones chirp faster so that they coordinate their hatching times. Mallard ducklings-to-be still in the egg can tell their mother to turn them if they are too warm or cool on one side. The ability of young and parents to recognize each other by calls seems to be related to the sociality of the species. The more social interactions that occur among the members of the group, the more it is necessary for communication to develop among individuals. Young birds in nesting colonies, such as guillemots, penguin chicks, and Cliff Swallow nestlings recognize the calls of their parents. And the degree of colonization (bigger, more closely packed) is related to the degree of recognition. In non-colonial birds no such recognition occurs.

Birds can often distinguish the songs of neighbors, mates, and strangers, and some bird species can recognize specific individuals. Studies of White-crowned Sparrows indicate that each individual voice is sufficiently different that birds are able to recognize one another. They need to hear only the first three notes of another's song; a difference in pitch distinguishes the bird. Many birds also exhibit varying degrees of vocal mimicry and imitate call notes or songs of other species. Scrub, Blue, and Steller's Jays can imitate a Red-tailed Hawk either to warn others that a hawk is present or to scare nesting birds away from their nest so that the jay itself can eat their eggs. Mockingbirds and others in the family Mimidae are well known for imitating other birds' calls although the main reason seems to be for a male to impress a female with its repertoire, which might exceed 200 different songs.

Non-vocal Communication

Although birds are primarily vocal, those with minimal oral abilities have developed other sounds. Kiwis stamp their feet when annoyed; Boat-billed Herons, storks, and albatrosses rattle or clap their bills; and the Ruffed Grouse "drums" with its wings. The male Common Nighthawk, to entice females, dives through the air and while pulling up, spreads his wings so that the feathers vibrate in the breeze much like the reed of a woodwind, making a noise commonly described as booming. The bird was once called a bullbat because its flight is reminiscent of a bat and its wing sounds boom like that of a bull. Male hummingbirds behave similarly; they dive and the feathers of their spread tail flutter and produce chirping sounds, different for each species of hummingbird.

Woodpeckers have calls and chatters, but they also advertise by drumming on anything that will produce a sound—dead trees, fence posts, houses, traffic signs, utility poles. Each species of woodpecker has its own drumming tempo and rhythm; I've watched Northern Flickers banging away at metal transmission towers, making an impressive racket. Both males and females drum as a form of communication and territory establishment, although males do it more frequently.

A rare ability in birds, echolocation—the emission of high-pitched sounds whose reflection indicates the direction and distance of objects—has been discovered in 16 species, notably the Edible-nest and Black-nest Swiftlets and Oilbirds. (The cave-nesting swiftlets of Asia are known for their contribution to "bird's nest soup," an expensive delicacy in China at $2,500 a kilogram. The nests, held together by saliva, form a gelatin when dissolved in water, producing the soup.) The swiftlets feed on flying insects by sight but nest in dark caves and have evolved a form of echolocation to find their way around. Unlike bats, however, the clicks they emit are audible

The Red-bellied Woodpecker employs a simple drum roll at about 19 beats per second. The bird was named for the reddish wash on the belly, though it is often not visible.

to humans as well as to other birds in the cave. Oilbirds belong to an unusual group of birds found all over the world (order Caprimulgiformes) variously called nightjars, goatsuckers, potoos, and nighthawks. Oilbirds of northern South America got their name from the native Venezuelan practice of collecting and killing the young birds and using their oil for cooking. Like the swiftlets, oilbirds are colonial cave nesters. They leave the cave at night to feed on fruit (the only nocturnal birds to do so) and use echolocation, audible to humans, to find their way; they also emit screeches that appear to be communication between individuals as a response to disturbance.

There seem to be no special adaptations to the ears of these birds to perceive echolocation.

SMELL: FIND FOOD, HOME, AND MATES, AND IDENTIFY PREDATORS

For many years birds were considered to be poor smellers but recent evidence shows clearly that some birds use smell to find food, communicate, and navigate. The dinosaurian ancestors of birds apparently had a pretty decent sense of smell, as judged by the size of their olfactory bulb (the part of the brain that processes smell), which was about 30 percent the size of the cerebrum (cognitive part of brain). The olfactory:cerebrum ratio of modern birds averages 20 percent although a few, like the kiwi, come in at 30 percent. The olfactory bulbs of small forest-dwelling songbirds comprise only 3 percent of the brain whereas those of some seabirds take up to 37 percent. Having a good nose for food is an essential survival skill for ocean-living birds. I have been on a few pelagic (from the Greek *pelagios*, of the sea) birding trips off the coasts of California, Connecticut, and New Zealand. These daylong boating expeditions are great for spotting seabirds that we can't see otherwise. On one journey, armed with saltine crackers, Dramamine, and image-stabilizing binoculars, I saw nothing unusual for the first hour or two as we motored out of sight of land. Then the bird guides began heaving buckets of buttered popcorn over the transom and birds started arriving almost immediately. Albatrosses can apparently smell food from as far as 12 miles away. I even heard a report of a Black-footed Albatross attracted by bacon drippings from as far as 20 miles away.

Smelling is also important for seabirds in orientation, navigation, and recognition. The Leach's Storm Petrel, a small seabird, spends most of the year at sea, returning to land only to breed. It digs burrows in the soil with its bill and feet and lays one egg inside; about 42 days later it hatches. After hatching, the parents spend the day at sea feeding, returning to the burrow at night to avoid predators and then regurgitate the victuals into the young's gaping mouth. To find the correct burrow among hundreds, the parent bird approaches the colony downwind and walks upwind, following the scent that will take it to its nestling. The importance of olfaction in finding young was demonstrated by experimentally plugging the nostrils of adult nesting prions (a type of small petrel) and Wilson's Storm Petrels. As a result, the birds were disoriented and had difficulty finding their burrows. Larger seabirds, less at risk for predation, return to their burrows during the day without olfactory clues; experimental blocking of their nostrils had no effect on navigation to their nest site.

Next time you visit the ocean shore, watch the birds sitting on the water, most likely gulls. Every once in a while, one will move its head as if it were sneezing. What the bird is doing is ejecting salt out of its external nares, openings in the bill like the nostrils of our nose. To smell, birds inhale air through their external nares; some birds also possess nasal glands in a depression in or on top of their eye sockets to remove salt from ingested seawater. Gulls, terns, shorebirds, some ducks, and all ocean-dwelling birds have these salt glands. The tubenoses, such as albatrosses, petrels, and shearwaters have a tubular structure on their bill used for olfaction and filtering saltwater. In a study of several dozen hawks and eagles, researchers found that these birds

Southern Royal Albatross showing its "tube nose."

also excreted salt solutions from their nares. Although they do not excrete as much salt as equivalent-sized seabirds, ridding themselves of salt ingested with their prey is important for water conservation.

A number of land birds are able to distinguish scents. When I was director of a field station in the mountains of northeastern California for several years, one of my non-academic (and less-than-pleasant) duties was cleaning the grease trap outside of the kitchen. The job entailed uncovering the cement-lined pit full of waste kitchen water, skimming the solidified fat off the surface and tossing it on the ground nearby knowing that porcupines, coyotes, and other creatures would devour it overnight. Within minutes, chickadees, juncos, nuthatches, and jays were picking away at the unappealing puddle of grayish sludge. Surely they were attracted by the odor.

In 1826, J. J. Audubon, artist and naturalist, conducted a couple of crude experiments to test whether vultures could smell. He put a stuffed deer complete with artificial eyes out in the field; the vultures shortly found the deer and shredded it. Then he placed some decaying carcasses of hogs outdoors, some uncovered and others swaddled with burlap, and noticed that the birds only found the uncovered carcasses. Audubon concluded that vultures cannot smell and that they find their carrion by sight alone. This misconception persisted until the 1960s when an ornithologist demonstrated that Turkey Vultures can find carcasses by scent, later identified as ethyl mercaptan, exuded by decaying bodies. Smelling skunkish, ethyl mercaptan is put in gas pipelines to detect cracks in the pipeline because it is harmless but detectable to humans at low concentrations. Although pipeline workers use more sophisticated ways to detect gas leaks, they also keep an eye out for vultures, as the birds will circle over any pipe cracks from which gas is seeping out.

The kiwi is among the few terrestrial birds whose lifestyle is largely dependent on olfaction. Given their nocturnal habits and flightlessness, one would think that large eyes would be advantageous. But in a sort of regressive evolution, the kiwi developed an enhanced sense of smell while its eyes and visual center of the brain diminished. They are the only bird with nostrils located at the tip of the bill, which helps them find food by smell and touch while probing in the soil for earthworms. Kiwis also react to the feces of other kiwis, which are often placed in conspicuous places such as the top of a log or root,

The rare Little Spotted (Grey) Kiwi exists in a few small populations in New Zealand.

by sniffing rapidly, sticking their beak into the air, and moving their head back and forth, similar to a mammal's response to an odor. This seems to indicate that kiwis use feces for social signaling, perhaps to declare a territory.

Finding Mates

Experiments with seabirds indicate that some birds recognize each other by smell well enough to determine genetic relationships and avoid inbreeding. Hector Douglas, a doctoral student at the University of Alaska, studied Crested Auklets in the Bering Sea and found that both sexes produce a tangerine-scented chemical, the concentration being highest during breeding season. The chemical acts as a tick and lice repellent; females prefer males with the strongest and perhaps the healthiest smell.

In another experiment, D. J. Whittaker of Michigan State University and several colleagues captured Dark-eyed Juncos from two different populations in California. One population inhabited an urban setting and the other a nearby montane region. The birds were housed in cages in identical environments and fed the same food for 10 months. Gas chromatography determined that the chemical odors emitted by the preen glands of each group were different. Differences between the sexes were found as well. This indicates a genetic basis for odor, suggesting that mate selection or species isolation (keeping different species from interbreeding) might be partially based on smell.

European Starlings incorporate strong-smelling herbaceous plants into their nests. The birds spend a considerable amount of time searching for the plants and it appears that the starlings choose the plants based on scent rather than appearance. The bacterial count in nests with green herbs is less than that in nests without herbs and the nestlings in herb-containing nests are healthier (as measured by blood cell counts). Not surprisingly, starling females prefer to mate with males with green herbs in their nests.

Scent as a Defense

The ability to detect the smell of a predator helps birds enhance their survivability. Experimenters in Spain placed the scent of a weasel on an active Blue Tit nest box when the young birds were eight days old. Video recordings revealed that when the parent birds arrived at the nest with food for the young they hesitated quite a while before entering. The parents did not make any fewer visits to the nest box because of the odor, but they spent less time there, reducing their chances of encountering a predator. Sleeping Great Tits exposed to the scent of a predator did not wake up or change their metabolism. They used anti-predator strategies such as roosting high in a tree or in a cavity before going to sleep rather than relying on olfaction.

Birds can also use scent as a defense against predators. The Northern Fulmar, a gull-like bird related to petrels that is commonly seen following fishing vessels, has a defense mechanism known for its "ick factor." If a predatory bird approaches, the fulmar vomits a stream of orange-colored liquid up to 10 feet, covering the intruder. Smelling like decaying fish, the liquid coats the potential predator's feathers, reducing their waterproofing properties, and making it difficult or impossible to fly. (Fittingly, "fulmar" comes from the Old Norse, meaning foul gull.) The Southern Giant Petrel has a similar defense mechanism, ejecting oil from its gut when threatened. Eurasian

Roller nestlings regurgitate and cover themselves with a foul-smelling orange liquid when frightened, not only becoming less palatable but also warning the parents returning to the nest that a predator was near. Northern Shoveler, Mallard, and Common Eider ducks eject feces onto their eggs when frightened by an approaching predator (or researcher). Some have speculated that the feces odor deters predators but it may simply be that the sitting birds eject feces as they take off.

A common misconception is that parent birds will abandon their nest if a person touches the nest, young, or eggs, leaving a scent. I've heard and read this hundreds of times and even a nature center website stated that parent birds will kill their young if humans touch them. I did years of research on blackbirds, grebes, grackles, cormorants, gulls, and ducks, counting nests and eggs. I have climbed Osprey nests and weighed and banded the young while the parents flew overhead and complained. But I was never aware of any nest being abandoned because of my actions. Parent birds, once they have invested the energy to lay and incubate eggs and especially after hatching them, are unlikely to give everything up and fly away because of a minor human disturbance. Like any animal (or person) in a desperate situation, birds will persist to the last minute. Survival depends upon it.

FLAVOR, BITTERNESS, SUGAR, AND SPICE: WHAT DO BIRDS TASTE?

From carrion-seekers to berry-eaters, the sense of taste varies greatly in the avian world with some birds rejecting distasteful fruits and others devouring hot peppers. A good deal of information is still yet to be uncovered and elucidated about the sense of taste in birds. Experimenting with birds and various substances in the lab tells us something, but we need more information from ecological and behavioral observations of birds in the wild to complete the picture. Compared to other vertebrates, birds have few taste buds. Catfish have around 100,000, rabbits 17,000, humans 10,000, and cats and lizards 500. Among birds, European Starlings have 200, the Japanese Quail 62, Bullfinches 46, and the barnyard chicken only 24. The taste buds of chickens are located so far back on the tongue that by the time the chicken tastes what it has eaten, it is swallowed.

Numerous examples have shown that birds can taste at least some flavors. Rice farmers lose millions of dollars each year in lost crops and expenses incurred trying to control pest birds. Dozens of techniques are employed, including flags, noise cannons, electric wires and fences, and model and real airplanes, to name a few. One of my former students used to ride shotgun (literally) in a cloth-winged biplane, shooting at blackbirds out a window. He moved on to another occupation after he blew a hole in one wing of the plane. Chemical deterrents to make the rice taste bad have been tried but they have proven expensive, ineffective, harmful to the birds, or not safe for food crops. One chemical that has seen moderate success is methyl anthranilate, which has been used to protect rice, fruit, and corn crops against some avian pests, and golf courses against the ravages of Canada Geese.

Lincoln and Jane Brower, along with their colleagues and students, studied various examples of plant chemical defenses. Their classic 1971 study of birds being stymied by bad taste is that of Blue Jays eating a Monarch butterfly and vomiting shortly afterward. Monarch caterpillars feed on milkweed and ingest cardiac glycosides. A Blue Jay chomping down on a caterpillar gets a nasty taste in its mouth almost immediately and the encounter is so intense that the bird learns not to eat Monarchs or the look-alike Viceroy. Recent studies indicate that Viceroy caterpillars accumulate salicylic acid from eating willow and poplar leaves and that they may be unpalatable on their own so perhaps the Monarch and Viceroy are co-mimics, both protected from bird predation. Clearly birds can taste and learn from the experience.

In 2004 the genome of the chicken was sequenced and researchers discovered that chickens lack the gene that allows the tasting of sweet substances. Another dozen or so birds tested also lack the gene so it's probably the case for many if not most birds. Red-winged Blackbirds will even choose plain water over a weak sugar solution. On the other hand, sugar lovers include parrots, hummingbirds, and other nectar and fruit feeders, even though they don't perceive the taste as sweet. Hummingbirds easily detect differences between the content of high and low sugar solutions and the amount of time hummers spend feeding is based upon the concentration of the solutions; higher sugar levels mean less time feeding. Douglas Levey of the University of Florida demonstrated that tropical tanagers can distinguish between solutions of 8, 10, and 12 percent sugar and that they prefer the latter, but

South American manakins did not detect any differences. The explanation is that tanagers crush the fruit as they eat it and manikins, like the Araripe Manakin, swallow it whole, so the tanagers are exposed to the chemicals of the fruit and manikins are not, at least not very much. The Long-tailed Manakin eats unripe fruits when nothing else is available but only enough to maintain its weight. The message here is that being unfussy and eating every available fruit, even those low in nutritive value, is a successful survival mechanism for birds that gobble bad-tasting fruits whole. This gives them access to fruits that other birds generally ignore and will only eat in times of food shortage.

Bitterness is common in toxic plants and is often a signal for birds to avoid a plant. For example, birds avoid foods high in tannins, an important chemical component of plant defense, especially in oaks. Tannins reduce the digestibility of proteins and are sometimes toxic in high concentrations. However, Blue Jays can subsist on a largely acorn diet for the winter if they have access to supplementary protein such as acorn weevil larvae. A species of aloe plant in South Africa produces dark nectar with a bitter taste; the plant is rejected by nectar-feeding bees and sunbirds, but is fed upon

Not much is known about the frugivorous Araripe Manakin, a rare endemic bird of eastern Brazil, only discovered in 1998.

by bulbuls, white-eyes, chats, and others that are apparently unaffected by the nectar's taste. The plant has apparently evolved to allow only a narrow selection of birds to distribute its pollen. Variation in the sensitivity to bitter tastes among different species is to be expected, but recent studies of the White-throated Sparrow indicate that there may also be differences among individuals within a species as their genetic makeup encodes for 18 different bitter taste receptors.

Chile peppers contain the chemical capsaicin, which, as you know if you have eaten one, causes mammalian pain receptors to produce mild to painful burning sensations. Capsaicin is such a powerful chemical that it is used in bear repellent at a 1–2 percent concentration; mail carriers use a weaker solution (0.35 percent) to deter dogs. Capsaicin has no smell or taste; its only stimulus is pain. Mammals learn to avoid eating peppers with capsaicin in concentrations as low as 100–1000 ppm (parts per million) but birds can eat peppers with concentrations up to 200,000 ppm as they lack the chemical receptor that reacts to capsaicin. This evolutionary strategy assures that pepper plants will not be eaten by mammals—which would chew and destroy the seeds in their guts—but by birds, which, for the most part, will pass the seeds through their digestive system unscathed and disperse them.

TOUCHY, FEELY: A LESS-EXPLORED SENSE

We know more about the touch sensations of mammals than we do about how birds respond to tactile stimuli. A bird's body is mostly covered with feathers that are insensitive to touch, although receptors are present in the skin at the base of the feathers. The scaly feet and smooth ramphotheca are alive but do not appear to be very good touch receptors. Birds have four kinds of tactile receptors: Herbst's corpuscles, Grandry's corpuscles, thermoreceptors, and nociceptors.

HERBST'S CORPUSCLES are the most widely distributed of the tactile receptors, found in the feathered skin, beak, tongue, legs, and feet. The receptors in the skin are associated with feather follicles and detect plumage disarray, causing the bird to commence preening or alter its flight behavior. Shorebirds, waterfowl, kiwis, parrots, ibises, spoonbills, and Whimbrels have abundant Herbst's corpuscles in their bills. Probing in the sand produces a pressure gradient; an object such as a mussel interferes with the normal gradient by blocking the water flow, which the Herbst's corpuscles detect, informing the bird of the mussel's presence. Herbst's corpuscles can also detect the movement of invertebrate prey burrowing through the substrate. The bill-tip organs of ibises function similarly to shorebirds, but the more aquatic ibises have a denser concentration of corpuscles than the other species, suggesting that these birds might be able to detect prey items in the water as well as below it. Rock Doves have a string of Herbst's corpuscles about 40 mm long. Anatomical, electrophysiological, and behavioral studies indicate that the pigeon can detect vibrations in the environment via these nerve endings. No definitive reason has been established, but it is logical that the movement of a branch, perhaps a predator, could awaken a bird sleeping in a tree. Some have speculated that birds can sense high-frequency vibrations that precede an earthquake; possible, but it's not likely this system evolved to escape earthquakes.

Whimbrels and other shorebirds use Herbst's corpuscles to forage nocturnally as well as during the day. The corpuscles allow the birds more time to feed and help them avoid diurnal predators and coordinate their feeding times with the tidal schedule.

GRANDRY'S CORPUSCLES are nerve endings found in the bill tip of aquatic birds; these sense organs, along with Herbst's corpuscles, detect bill tip movement. Kiwis have both Herbst's and Grandry's corpuscles. The bill-tip organs of three families (sandpipers, kiwis, and ibises) appear to have evolved independently, so it may be that other probing birds have similar organs that we have yet to discover.

TEMPERATURE RECEPTORS or THERMORECEPTORS consist of free nerve endings in the skin but are primarily located in the beak and tongue. Cold thermoreceptors are more abundant than heat receptors and various areas of the feather-covered skin are differentially sensitive to temperature. The skin on the back of the Rock Dove is more sensitive to heat than the skin on the wings and breast, possibly because the back of the bird is more often exposed to the sun than other parts of the body. Unfeathered skin areas such as the legs and feet are fairly insensitive to either warm or cold input. The brain also contains thermoreceptors, which signal the bird to make appropriate physiological or behavioral adjustments to maintain the proper body temperature.

NOCICEPTORS are free nerve endings that are receptive to any stimulus that threatens or causes pain. They are located in both the skin and the beak and are sensitive to strong mechanical forces, chemical irritants such as toxins in plants, and heat over about 113°F. Birds respond with increasing blood pressure and heart and respiratory rates.

The more advanced our technology becomes, the more we discover about the senses of birds. It has been only since the 1980s that we have begun to understand that birds can see ultraviolet, their hearing does not diminish because they replace damaged hair cells in the ear, they detect objects by differences in pressure, and they can eat hot peppers because they simply can't taste them. What else do we have yet to learn?

UPS AND DOWNS

The Animals That Conquered the Air

A bird maintains itself in the air by balancing, when near to the mountains or lofty ocean crags; it does this by means of the curves of the winds which as they strike against these projections, being forced to preserve their first impetus bend their straight course towards the sky with divers revolutions, at the beginning of which the birds come to a stop with their wings open, receiving underneath themselves the continual buffetings of the reflex courses of the winds.

—LEONARDO DA VINCI, "Flight,"
The Notebooks of Leonardo da Vinci

Fish gotta swim and birds gotta fly, as the song goes. Without the power of flight, birds probably would not be here today in any significant numbers and certainly not 10,000 species all over the globe. After insects started flying about 350 million years ago, their evolution and success proliferated. Pterosaurs, the only flying reptiles, came into being about 200 million years ago, but how nimble they were is a still-unsettled question. Birds, with their light bodies, strong muscles, sturdy skeletons, and those marvelous feathers, conquered the air about 150 million years ago; bats took another 100 million years to come on the scene.

Watching birds from my kitchen window I marvel at how different they are even though they all have to meet the requirements of flight. The California Towhee scurries across the ground like a mouse, only occasionally hopping into the air; the Black Phoebe launches itself off one perch to land on another and then returns to the original one; the Yellow-rumped Warbler darts among the branches of the crabapple tree; and a crow hops from the top of one redwood and glides to the next. The Nuttall's Woodpecker flaps and swoops from one tree trunk to another. Swallows ply the open sky searching for insects while swans high above subtly move their wings as they migrate southward. Even with basically similar body forms and complements of feathers, birds have adapted a myriad of flight styles.

Flying is essential to survival for the vast majority of birds. It is not a skill that develops after years of practice, like our learning to play the violin or pole vault—it's an innate ability that allows birds to find food and nesting sites, escape predators, and avoid weather extremes, in order to exploit habitats and niches that no other animals can. As masters of the air, birds are found everywhere on the planet except the center of Antarctica. For the few flightless birds in existence, the loss of flight came about secondarily.

The flight of birds has always fascinated us. In mythology, Icarus's father, Daedalus, constructed wings of feathers bound with wax; the wax melted when Icarus flew too close to the sun and fell to the sea. The wings were of bird-like construction with a sequence of overlapping and increasingly long feathers from front to back and a curved upper surface, indicating some early knowledge of aerodynamics. In the 15th century Leonardo da Vinci designed the ornithopter to include a series of pulleys to enhance human power as he rightly deduced that human arms were not strong enough to flap extended wings. A locksmith in Sable, France, made an impressive attempt at flying in the 17th century. He fashioned a set of four wings, two on each side, front and back, and attached them to a rod over his shoulder. With some effort, he was able to flap himself to greater and greater heights, step by step, finally reaching rooftops from which he glided across a river. Over a century ago, a few eccentric inventors created machines that were designed to fly with wings powered by human arms. Since our pectoralis muscles are only 1 percent of our body weight, none of these machines got very far. For an average human adult male to fly, a wingspan of 22 feet would be required. The wings would simply be too long and heavy for him to lift. There have been hundreds of other attempts of human-powered flight and a few have been reasonably successful, like

A 17th-century painting depicts Icarus falling from the sky after getting too close to the sun.

the Daedalus aircraft that an Olympic cyclist piloted and powered for 37 miles in 1987.

In *Empire of the Air* (1881), Louis Pierre Mouillard, a French philosopher and glider inventor, described the flight of birds in an essay. "All my life I shall remember the first flight which I saw of the *Gyps fulvus*, the great tawny vultures of Africa. I was so impressed that all day long I could think of nothing else; and indeed there was good cause, for it was a practical, perfect demonstration of all my preconceived theories concerning the possibilities of artificial flight in a wind." Orville and Wilbur Wright observed birds too, especially vultures, moving their wings as they flew, and incorporated mechanisms to warp the wings of the powered Wright Flyer, which first took to the air in 1903. But even after the Wright brothers' flights, knowledge of how birds flew was minimal. In an amusing but rather embarrassing essay titled *Principles of Bird Flight* published by the New York Academy of Sciences in 1905, May Cline wrote that birds inhale deeply, making themselves lighter than air, and fly off. Sometimes they inhale too much and explode, hence the pile of feathers you occasionally see in the woods. We know a bit more about aerodynamics today.

On a visit to the Udvar-Hazy Center in Virginia, an annex of the Smithsonian National Air and Space Museum, I saw the space shuttle Discovery, a supersonic Concorde, the bomber Enola Gay that made Hiroshima famous, and all manner of other aircraft. Possessing single and double wings, straight and swept back wings, blunt and pointed noses, sleek and dumpy, they all flew. Without those air machines, we would live in a two-dimensional world because all we can do without them is move horizontally along the ground or water.

To understand a bird's world we need to know something about flight. I spent six months training for a pilot's certificate and flew small planes for several years before deciding that I had used up eight of my nine lives. General aviation is not particularly safe, but risk is part of the appeal, like skydiving or paragliding. But I didn't do it because it was risky; I did it because as an ornithologist teaching students how birds fly, I thought I should have some personal experience. That experience imparted some cold sweat, shaky legs, and nausea. Stalling, diving, making 60-degree turns, flying in turbulent air, maneuvering blind through clouds, and trying to land perpendicular to a stiff wind made me appreciate what birds have to deal with. Watch a flock of geese landing on a rainy, blustery day and you will see what I mean. How do birds do it?

ADAPTATIONS FOR FLYING: SKELETONS, MUSCLES, LIGHTNESS, AND POWER

Every adaptation that an organism possesses came about through evolution. Changes in genes or chromosomes that proffer an organism a bit of an advantage will be assimilated, built upon, and refined, typically in very small increments. The skeleton, muscles, physiology, and behavior of each bird were crafted by natural selection over a couple of hundred million years, resulting in these feathered masters of the air. So what do birds need to become creatures of the air? Besides feathers as airfoils, the requirements are a reduction in body weight and sufficient power and energy to move the wings.

The Skeleton: Let There Be Lightness

Being light makes getting into the air easier and reduces the metabolic cost of flying—and anything that reduces energy expenditure adds to the survival quotient. The largest flying bird that ever existed, *Argentavis magnificens*, lived about six million years ago in what is now Argentina. With a wingspan of about 21 feet and weighing more than 150 pounds, 16 times as heavy as a Bald Eagle, it flew, but just barely. It had to run downhill or launch itself from an elevated perch and depend on wind currents to keep it aloft, making it primarily a glider. The male Kori Bustard is the heaviest flying bird alive today at nearly 40 pounds but it prefers to walk. When threatened it will run and if the threat continues it will it fly—yet it is just too heavy to fly any distance. Light weight is critical to sustained flight: the heavier the bird, the more muscle power is required to lift it, which requires bigger wings to provide more lift, adding more weight and the demand for even more power.

Even some of the most ungainly appearing birds, such as the Greater Flamingo, have developed adaptations for flight.

To lose weight so birds could take to the air, the beak replaced heavy teeth and many bones fused together or were completely lost. Some thoracic (chest) vertebrae melded, eliminating some individual bones and stiffening the back to support the flight muscles. The lumbar, sacral, and some caudal vertebrae fused with the bones of the pelvis, forming a structure called the synsacrum. This construction stiffens the dorsal skeleton, strengthening the entire body to withstand the stresses of flight and landing. The rest of the caudal vertebrae fused into the pygostyle (Greek for rump pillar)—it is sometimes called the "pope's nose"—a short bony structure that is covered by muscle and skin and to which the tail feathers are attached.

With a rigid skeleton and wings that are pretty much restricted to locomotion, birds need flexible necks to move their head for essential everyday functions like feeding, eating, and nest construction. If you have ever seen an unresponsive bird that just collided with a window, what looks like a broken neck is simply a very flexible one with 13–25 cervical vertebrae that allow the head of the bird to droop. Mammals, with few exceptions, have only seven bones in their neck, and although some mammals have limited neck movement, others, like the giraffe with 10-inch-long cervical vertebrae, are amazingly flexible.

Some bones of the "hand" (outer part of the wing) have been lost or fused, leaving only three digits instead of the typical five found in most land vertebrates. The "thumb" supports the alula or bastard wing, a small but important structure in flight. A number of bones in the toes and legs are fused or have been lost. A bird actually walks on its toes and what appears to be a backward-facing knee is the metatarsus bone, essentially an elongated ankle. The major flight muscles, those providing power, are attached to a ventral extension of the sternum or breastbone, the carina (Latin for keel). When we carve a Thanksgiving turkey, we usually start by slicing pieces of the side of the breast, making a pile of white meat out of the two major flight muscles, eventually exposing the carina. The sternum is braced by the wishbone or furcula (Latin *furca*, fork). As the flight muscles compress the thorax during the downstroke, the furcula bends, absorbing some of the stress to the skeleton; on the upstroke, the furcula expands, helping a bit in moving the wings upward.

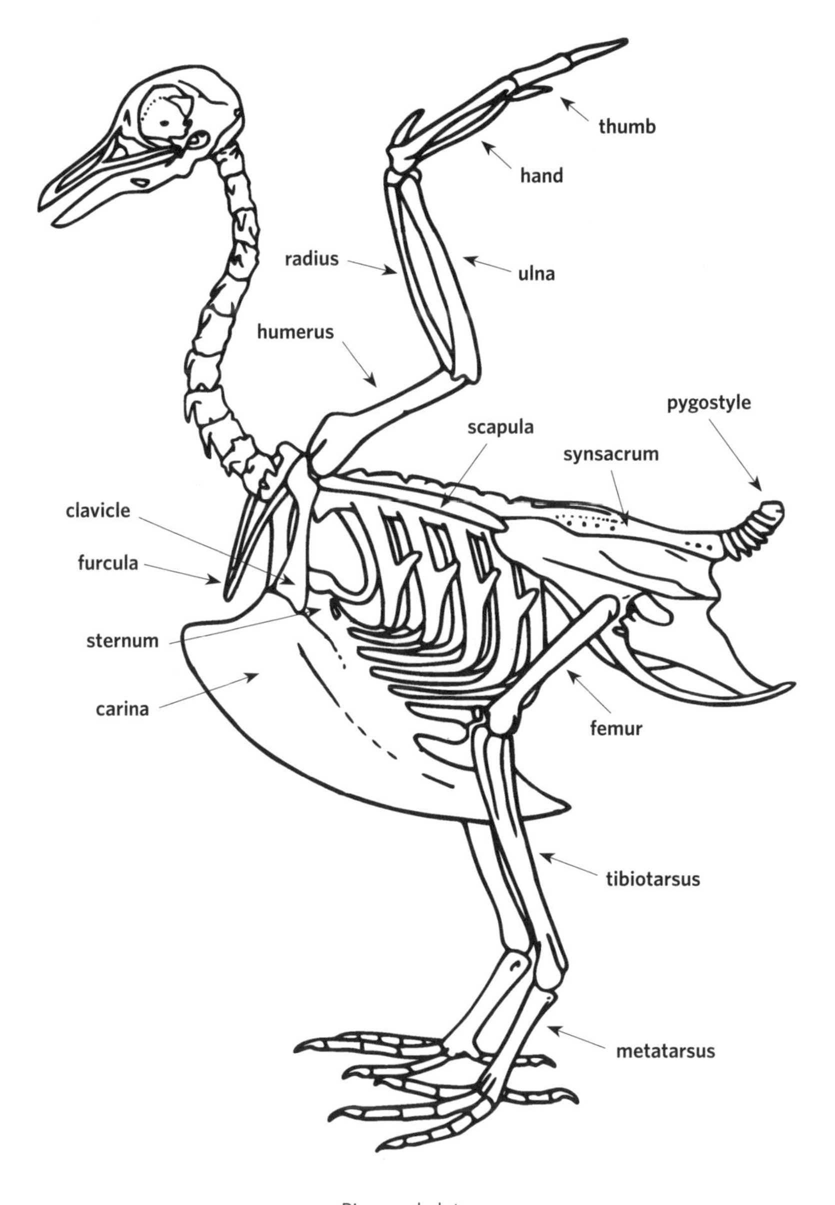

Pigeon skeleton.

Have you ever picked up a bird that had died recently? After the body dehydrates, a sparrow-sized bird is almost weightless. Many bird bones are mostly hollow, with struts and airspaces, like the triangular construction of a truss bridge. Physics tells us that a hollow cylinder is harder to bend or break than a solid tube of the same mass and material, and internal struts make the cylinder even stronger. The hollowness of the bone varies among birds: eagles, owls, swans, and cranes have thin-walled arm bones while seabirds, loons, and penguins and other divers have the thick-walled arm bones needed to withstand the stresses of underwater swimming. The hollowness, internal struts, density, and shape of bird bones, as well as the reduction in number, makes for a solid but light "fuselage."

Dropping weight is also achieved by the reduction of internal organs and seasonal physiological changes. Most female birds (with exceptions like the kiwi) have only a left ovary and oviduct, which shrink noticeably during the non-breeding season. The male's testes are tiny during the non-breeding season and may enlarge by as much as 300 times when the nesting season commences—at this point the birds are not migrating so they can afford a bit more weight. Except for the Ostrich, birds do not have a bladder. Carrying water only makes birds heavy, so instead of producing watery urine, their nitrogenous waste is in the form of a white paste of insoluble uric acid, which contains only about 5 percent water. When drinking nectar, hummingbirds take on a lot of water and thus weight. To reduce the amount of water they carry, they transpire some of it from their respiratory system—huffing and puffing 250 times per minute—and remove the rest via their very efficient kidneys. A typical hummingbird, in fact, will eliminate twice its weight in water every day.

Muscles and Physiology: The Engines for Flight

Being light is necessary but not sufficient. Although a bird can glide and soar in the wind, the range and duration of flight are limited—and then there's the problem of getting into the air in the first place. Survival in the sky requires the physical force of muscles and the physiological engine to power them. How else could the amazing Alpine Swift fly for 200 continuous days without landing?

Of the 45 muscles involved in powering and directing the flight of a bird, the two most important are those that comprise the breast. The breastbone of the bird holds the large pectoralis and the much smaller supracoracoideus underneath; together they account for about 17–30 percent of a bird's total weight. The pectoralis pulls the humerus down and forward for the power stroke. The supracoracoideus, via a tendinous pulley system through the shoulder, pulls the wing up and back in a recovery stroke.

The avian circulatory system, like that of mammals, is driven by a four-chambered heart. The avian heart accounts for about 1 percent of total body weight, twice that of a mammal's. Smaller birds have relatively larger hearts; a hummingbird's heart comprises about 2 percent of the bird's weight compared to a turkey's heart at less than 0.5 percent. (Data on bird heart weights were gathered by Frank Hartman in 1955 who shot and dissected more than 1340 birds of 291 species, the kind of research that is frowned upon these days.) Bird hearts pump a greater volume of blood per minute, and their heart rates tend to be faster than the hearts of similar-sized mammals. The Blue Jay's heart beats 165 times per minute, the American Robin's 550, the Blue-winged Teal's up to 1000, and some hummingbirds an astounding 1200. Small mammals like the house mouse have comparably high heart rates at 670 beats per minute versus a human's moderate 60–90.

Bird hearts are more efficient than mammalian hearts because the faster heart rate and the efficiency with which the ventricles fill and empty circulate more blood with each contraction. Avian heart cells are also stronger and more efficient at absorbing oxygen. Blood pressure is also generally higher in birds than mammals with the pigeon's measuring at 135/105 and the chicken at 180/160. The blood pressure of rats, domestic dogs, and humans are similar at 120/70; the relaxed guinea pig comes in at 80/55. Domestic turkeys, bred for large breast muscles, have very high blood pressure at 235/141, and squabbles among captive birds occasionally lead to heart attacks or aortic ruptures. Once, while banding a Northern Cardinal, I could feel its heart pulsing rapidly and strongly. Before I squeezed the band closed, the bird died, most likely of a heart-related issue. Wild birds have heart attacks, but cardiac problems are much more common in caged pet birds as they tend to be "perch potatoes" and are often fed junk food.

The strong circulatory system necessitates an effective respiratory system. In mammals a muscular diaphragm helps the lungs expand and contract to inhale and exhale air. Birds do not have a diaphragm and their lungs do not move. Extending from the lungs are nine (typically) thin-walled air sacs that, along with the lungs, comprise the respiratory system. The air sacs account for 15 percent of a bird's body volume, compared to 7 percent for mammalian lungs. Some air sacs penetrate hollow bones like the humerus and fill areas between the skin and muscle. During inspiration the chest expands, and the decreased pressure brings air in through the lungs into the air sacs and then back through the lungs again on expiration. The continual flow minimizes the mixing of new oxygen-rich air and stale carbon dioxide-saturated air as happens in mammalian lungs. Air sacs also serve as a cushion for the viscera upon landing, protection for birds diving from air into the water, and an adjustable buoyancy mechanism for many water birds. Air sacs have also been modified for courtship displays as in the Magnificent Frigatebird, Prairie Chicken, and Sage Grouse. Because birds lack sweat glands, the respiratory system also functions to eliminate heat from the body.

Many birds fly at high altitudes, especially on migration, but their high respiratory rate provides them with all the oxygen they need. Adult humans take 10–12 breaths per minute. An Ostrich's respiratory rate is five or six breaths per minute, a pigeon's respiratory rate is 28 breaths per minute, a House Sparrow's 57, and a European Robin's 97. The amazing Bar-headed

Male Magnificent Frigatebird displaying to females by expanding its air sacs.

Goose, which migrates over the Himalayas at almost 30,000 feet, has hemoglobin especially adapted to attract oxygen, important for survival at high altitudes where the oxygen levels are 35 percent lower than at sea level. Bar-headed Geese breathe more deeply and at a faster rate when more oxygen is needed; they can also hyperventilate, but without suffering the effects of lightheadedness that humans do.

Metabolism and Energy

Exercise physiologists have told us for years that a good workout regimen will lower our basal (resting) metabolism. But if we only sporadically exercise, that workout only increases our resting metabolism. Birds, on the other hand, after a long flight or just a short hop across the field, exhibit a reduction in their resting rate because their metabolism is so finely tuned. Although the basal metabolic rate of small birds is 40–70 percent higher than small mammals, the energy cost of activities above the basic level seems to be no more in birds than in mammals. In fact, flying is more energetically economical than walking, hopping, or running is for large mammals.

As an undergraduate I earned minimum wage by cleaning the bottom of birdcages and separating the poop from the leftover food. The researcher I worked for was trying to determine the energy requirements of molting birds in captivity by measuring their activity, amount of food eaten, and feces produced. This approach gave rough estimates of energy use but more accurate methods were developed later. In a classic (I'd even call it cool) experiment, Vance Tucker put oxygen masks on parakeets and trained them to fly in a wind tunnel. Graphing oxygen consumption against flight speed produced a U-shaped curve, indicating that at low and high speeds energy consumption is higher than at a moderate, supposedly optimum, flight speed. Although this idea has been debated since the 1970s, recent research indicates that Tucker was probably on the right track.

In the past ornithologists would estimate the time and energy budget of a bird in the field by watching and noting its behavior at 10-second intervals. So for each minute of observation one would have six data points describing the bird's actions—feeding, flying, perching, sleeping, chasing, or whatever. Five hours in the field would yield 1800 notations. The work was tedious and not particularly stimulating, but by combining those data with lab studies on how much energy birds use, we can derive an understanding of how much energy birds need to survive in the wild on a daily basis. With today's methodology and technology, including fitting free-living birds with electronic gear that transmits heart and respiratory rates, ornithologists have accumulated substantial data about the energy required for flight.

The energy needed depends on the size of the bird, wing shape, the mode of flight, and environmental conditions. To estimate energy needed for a particular activity, researchers first determine the basal metabolic rate (BMR), which is the energy used while resting in a thermoneutral environment (no extra energy used to cool or warm the bird). The flight energy requirements of hummingbirds, for example, are 12–15 times their basal metabolic rate, so they forage for nectar, which is high in sugar. John Videler's book, *Avian Flight*, reports energy used during flight as a multiple of BMR. Some examples: Barn Swallow 5.1, European Starling 10.3, Rock Dove 18.0, Red-footed Booby 2.7, and the Wandering Albatross—the most efficient flier—at 1.4.

FEATHER AND WING SPECIALIZATION

Several avian adaptations make flight possible, but feathers are truly indispensible. Feathers are made of the protein keratin (as are beaks, scales, and claws), rendering them light and soft but also strong and flexible. A unique feature of birds, feathers could not have come about *de novo* into their present form; they arose slowly from some simple basic structures into the array of forms we have today. *Archeopteryx*, long considered an intermediate between reptiles and modern birds, had wings with feathers very much like the flight feathers of today's birds. Debates arose about how these wings were used. Were they used for powered flight, gliding, or something else? Few scientists recognized that the feathers of *Archeopteryx* were pretty advanced and that natural selection apparently molded them many millions of years earlier from more primitive structures. Like most ornithologists, for many years I accepted the hypothesis that feathers evolved as an extending, flattening, and thinning of reptilian scales. This made sense since scales and feathers are made of similar material and birds evolved from reptiles. As dinosaurs, then mammals and birds, evolved, so did homeothermy (warm-bloodedness) and it made sense that elongated scales came along to provide insulation.

But Richard Prum, an ornithologist at Yale University, in a flash of insight, wondered about feather origins and eventually gathered enough evidence, with help from colleagues and students as far away as China, to determine that feathers were actually novel structures that evolved independently. The first clue to this revelation was the realization that scales develop as flat structures and feathers develop as curled tubes that unfold. The first feathers were little fuzz-like structures in shades of black and brown. To make a long story short, it appears that feathers evolved for visual communication, display—and that's what kept them moving on the evolutionary path toward longer and more refined structures that along the way became useful for insulation (as down feathers), gliding, and eventually flying.

Because of their ability to endow birds with flight, feathers have been imbued with magical and mystical powers over the ages, but birds are still subject to the laws of physics. The four forces of flight are lift, gravity, thrust, and drag. The wings and tail provide lift to counteract the force of gravity. When birds flap their wings in a downstroke, the wingtips move forward and down, producing forward thrust, sort of like humans swimming the crawl stroke. Thrust from the flapping wings propels the bird through the sky, counteracting gravity, producing lift, and overcoming drag due to the friction of air moving over the bird. Then the wings reverse in a recovery stroke and move upward and back while flexed inward to minimize drag.

When a bird's body accelerates forward, lift is proportional to thrust—the faster the bird flies, the more lift is generated. But if the bird angles upward to gain altitude, as it does on takeoff, for instance, the angle of the wing increases relative to the oncoming air. This is the "angle of attack." If the angle of attack is too great, the wings lose lift, causing a stall. The alula, the few feathers attached to the thumb, can rise and form a slot over the wing, smoothing the air and allow a higher angle of attack without a stall.

Lift and thrust move a bird through the air, but to steer, brake, and land birds have specialized contour feathers. Each contour feather has a shaft with parallel barbs emanating on both sides to form a flexible vane. The portion of shaft extending below the vane is the calamus or quill. You are

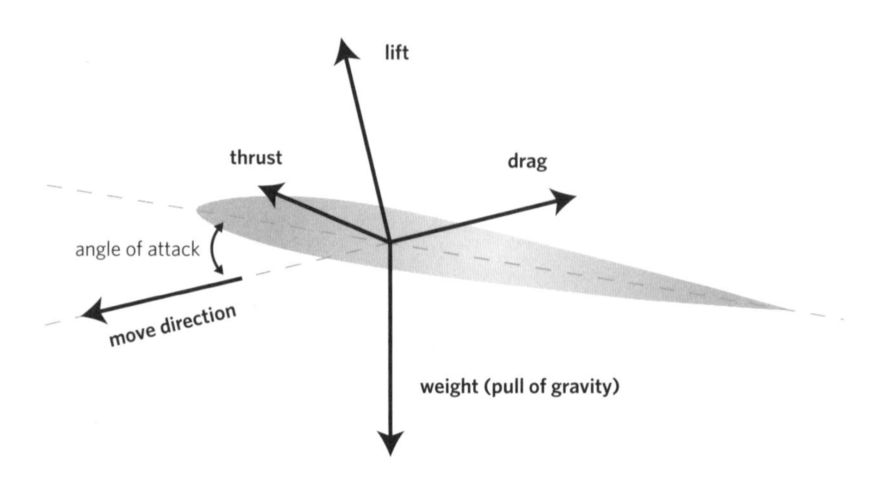

Forces of flight.

probably familiar with how these feathers feel: you can run your index finger and thumb down to separate the barbs and make the feather ragged; just as quickly, like Velcro, if you run your fingers upward along the vanes the feather barbs can be neatly reattached.

Different contour feathers perform different flight functions. Producing thrust, the 9–11 primary feathers attached to the hand have different-sized vanes on each side, making them asymmetrical; they can be individually rotated to reduce drag and turbulence by forming horizontal slots between them, aided by notches near the ends of the feathers. The secondary feathers of the arm, 9–11 in songbirds and up to 40 in albatrosses, are symmetrical. They move in unison because they are connected across their bases by a ligament that allows them to be moved simultaneously to provide lift or help in landing. Three to five flight feathers are attached to the thumb, forming the alula, which helps to reduce turbulence over the wing when the alula is lifted to form a slot.

Particularly important are the primary and secondary wing coverts (think "cover")—the feathers on the leading edge of the wing that smooth airflow by forming a tapered surface to cover gaps between the bases of flight feathers. The feathers of the shoulder, the tertials and scapulars, fill in aerodynamic

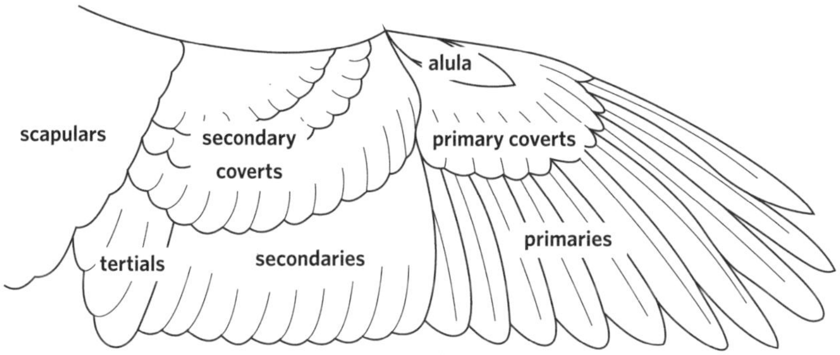

Bird wing indicating contour feather types.

gaps and help with lift. There are covert feathers over much of the body including the tail, filling in gaps to form an aerodynamic shape.

Attached to the muscles supported by the bony pygostyle are the tail feathers, which spread, fold, and move (up, down, and laterally) to enable lift, turning, and braking. Cats and other predators that attack birds from behind to avoid detection will often end up with a mouthful of feathers as birds occasionally drop their tail feathers in a "fright molt" to distract a predator. But trading tail feathers for escape to safety means that the bird's life will be more difficult until the feathers grow back in six weeks or so. The bird has to fly faster to compensate for the loss of lift and the aid of the tail in turning and braking, and the wings must work harder.

The color of some feathers plays a surprisingly important role in flight. The pigment melanin, which produces brown and black colors, increases the resistance of feathers against abrasion from the turbulent air around the wingtips, airborne particles, the wear caused by feathers rubbing against each other, and degradation by feather bacteria. This is why many large birds (Snow Geese, White Pelicans, the European White Stork, and some gulls and terns) have black or dark wingtips, or totally black wings (like frigatebirds and murres).

Configuration of Wings

When leading bird hikes, I advise participants to focus not on color for identification, but on the bird's shape. In 1934, Roger Tory Peterson, a well-known ornithologist and artist, published a field guide to birds and included on the inside covers (a format that would became the model for subsequent guides) silhouettes of birds perched and flying because shapes are so important in identification. A significant part of a bird's silhouette, especially in flight, is its wings. Many variations in wings have arisen as different habitats have molded birds for different niches, each requiring its own mode of flight. Short and wide, long and narrow, and all manner of intermediate wing shapes have evolved, including those for underwater "flight." Like the beak, the wings are elegantly suited to carry on the lifestyle of a particular bird. But, as in airplanes, the laws of physics circumscribe the limits of wing variation.

Two measures of a bird's wing, the length and the width, tell us a lot about the bird's flying style. The wing aspect ratio is the ratio of the length of the wing to its width. Wing loading is the relationship of weight to wing area; the less weight a wing has to support, the less energy a bird has to use. Soaring and gliding birds have bigger wings and thus have smaller wing loadings than songbirds, so they need less power to generate lift and can soar and glide without using much energy. Birds like chickadees and bullfinches cannot soar and can glide only for short distances because their wings are too small to produce sufficient lift without considerable power. Most bird wings can be categorized as one of four types: high aspect ratio wings; high lift, low aspect ratio wings; high speed wings; or elliptical wings.

HIGH ASPECT RATIO WINGS are much longer than wide and are used for slower flight as seen in storks, eagles, frigatebirds, pelicans, albatrosses, and seabirds that frequently hover or soar. These wings have deeply notched primary feathers with narrow, pointed wingtips to minimize turbulence and enable long periods of soaring, like the weeklong flights of frigatebirds. Like a tightrope walker with a balancing pole, long wings provide stability and are less susceptible to jostling by air currents. On the other hand, they cannot be used for quick maneuvers. Big birds with high aspect ratio wings usually need to take off from an elevated perch or into the wind. A stiff wind is helpful for landing, which could otherwise be difficult. Military servicemen stationed on Pacific islands during World War II gave the nickname "gooney bird" to the albatross after watching these birds make ungainly and often clumsy takeoffs and landings.

HIGH LIFT, LOW ASPECT RATIO WINGS are broad (wider but shorter) wings found on many raptors like hawks, eagles, vultures, and the Osprey. The birds can soar, glide, and maneuver well, but without much speed. High lift wings don't have the stability of high aspect ratio wings and they produce more drag, but they provide enough power to soar on thermal updrafts and lift heavy prey. Slow flight is important to predatory birds and scavengers because it allows the birds to scan for potential food sources. The feathers of these wings are curved in cross section to provide lift and the slots formed at the tips of the wings by the primary feathers help to reduce wing tip turbulence (vortices).

HIGH SPEED WINGS are long and pointed and common in ducks, falcons, auks, as well as birds that feed aerially like swifts, undergo long migrations such as shorebirds, or "fly" underwater like puffins and murres. The wings taper to a point to minimize drag and energy consumption but require fast wingbeats to produce lift. Slow flight is difficult.

ELLIPTICAL WINGS are short and rounded to allow for quick maneuvering in tight spaces like forest and scrub habitats. Crows and many other songbirds have elliptical wings for quick changes of direction and minimal drag. Their wingbeat is typically rapid and the primary feathers are strongly slotted to prevent stalling during quick turns and frequent landings and

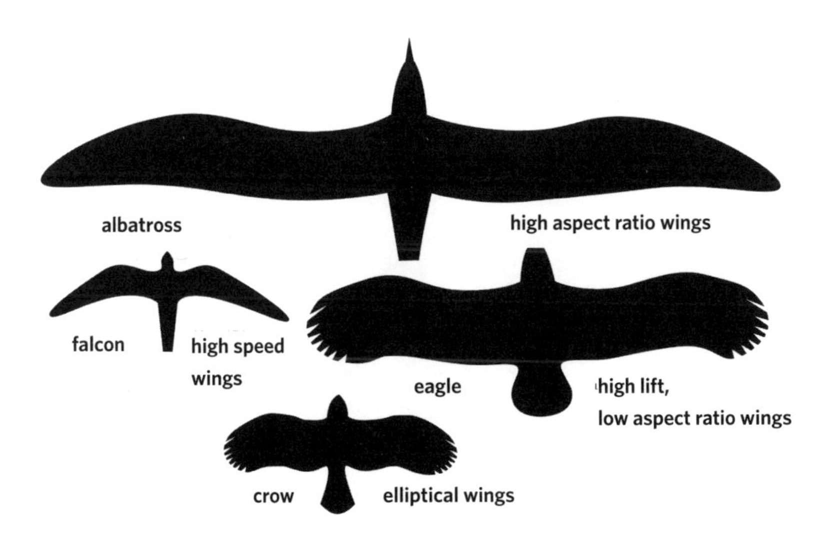

Wing aspect ratios.

takeoffs. The Cooper's Hawk pursues its mostly avian prey with a quick burst of speed and agile avoidance of tree branches and shrubbery. The tradeoff for agility is high energy use.

Tundra Swans have a wingspan of nearly six feet, which provides the birds with considerable lift. They slide smoothly through the air using almost imperceptible movements of their primary feathers. Canada Geese flap a little harder and faster as they have smaller wings, but again mostly with their primary feathers. Ducks, especially fast-flying teal with smaller wings, need to flap rapidly and use most of their wing surface for lift and thrust. The small wing area of many songbirds provides minimal lift, so they have to flap even faster and harder to stay aloft, at least 11 mph, perhaps all the way from the tundra.

As a bird flies, it adjusts its speed and altitude by flapping faster or slower. To change direction the bird adjusts the lift of its wings. If the left wing is pulled in toward the body, lift will decrease on the left side. The fully extended right wing is producing more lift so the bird tips and turns to the left. However, this causes a "skid" in the air—think about making a too fast

Equipped with long and wide wings, Tundra Swans merely need to move their wingtips to almost effortlessly ply the skies.

left turn in your car. To minimize skidding, the tail twists in the direction of the turn, like an airplane rudder. To lose altitude, both wings are pulled inward. All this is difficult to discern unless you watch a slow flying bird to see the constant movements of its wings and tail.

Birds have sophisticated ways to deal with the vagaries of air currents. On a blustery day, flying can be a real challenge. In Wyoming (where a wind gauge consists of a bowling ball on a chain, 45 degrees indicating a light breeze) I've watched Tree Swallows trying to catch insects in a gusty wind. They dipped, bobbed, dove, and turned so quickly that it was impossible to determine exactly what adjustments the birds were making. A wind tunnel experiment on Barn Swallows provides some explanation. The air in the tunnel was kept constant (no turbulence) while the researchers measured the birds' flight responses between 8 and 33 mph. Wingbeat measurements showed that their wings beat faster at high and low speeds than at intermediate speeds. Wingbeat amplitude (the distance the wings moved above and below the body) increased with flight speed. Wingspan, the width of the tail spread, and the tilt of the body all decreased as the birds flew faster. What the birds were doing was increasing lift at low speeds and decreasing drag at high speeds. It is hard to imagine how accurately and quickly those Tree Swallows were able to adjust in winds that continuously changed both speed and direction.

Hummingbirds do something a bit different. While most birds get thrust from the downstroke and then bend their wrists back to pull the wing up in a recovery stroke, hummingbirds rotate their wrist after the forward stroke to power the backstroke which also produces lift. Imitate a flying goose or gull by flapping your arms forward and down and then up and back, sort of like swimming. To imitate a hummingbird, hold your upper arms against your side and flap your lower arms the same way. Since the hand and primary feathers comprise 80 percent of a hummingbird's wing, it is not surprising that they do most of the work. The pectoralis and supracoracoideus muscles of a hummingbird are about equal in size, compared to the 80:20 percent ratio in other birds, and the weight of these two muscles comprises 30 percent of the hummingbird's total weight, almost twice that of most birds.

TAKING OFF, LANDING, SOARING, AND GLIDING

Before a bird can flap, soar, or glide, it has to get itself into the air. What steps got birds into the air? Paleontologist Xing Xu and his colleagues in Beijing reported that *Microraptor*, a four-winged dinosaur of the Cretaceous period about 125 million years ago, could glide, representing an intermediate step to the flapping flight stage. It had long flight feathers both on the legs and wings, a fan-shaped tail, and light and dark bands of proto-feathers providing coloration, including iridescent black.

For years two competing theories explained how birds became airborne. The arboreal theory proposed that the birds climbed into trees and glided down. Some evidence supports this theory. Models of *Microraptor* were put in a wind tunnel at MIT and the lift and drag forces measured. With all four limbs extended horizontally, the model was an excellent glider. The opposing cursorial theory says that birds ran along the ground and eventually generated enough speed to produce lift for takeoff; evidence for this idea is weak. In 2003, a third idea came into play: birds used their wings to climb over rocks or scale walls, flapping and running over steeper and steeper

Microraptor: speculative illustration of the animal in flight.

inclines until they were airborne. Using Chukar Partridges, researchers looked at the development of the ability of newly hatched birds to scale an incline, from crawling on all fours to using symmetric wingbeats to fly by 20 days after hatching. They speculate that the development of flight in these partridges mirrors the evolutionary steps to the power of flight, but the debate continues.

Some birds jump off cliffs or tree branches, run into the wind, or run across the top of the water, but a number of birds just launch themselves into the air from wherever they are. Go to the nearest pond or lake and look for a Mallard, the most common duck in the Northern Hemisphere. It has a rather stout body and average wing dimensions, but watch it take off at a 60-degree angle! How can that be? First, it spreads its wings fully and flaps downward, the wrist (between the primary and secondary feathers) hitting the water to lift itself off the surface to become airborne. If a bird or airplane is aimed too steeply upward, the smooth flow of air over the wing is disturbed, lift is diminished, and the bird (or airplane) stalls and falls downward. The Mallard is able to take off steeply but avoid a stall by extending its alula, the three to five feathers attached to the thumb at the wrist. The alula lifts up like slats on an airplane wing and smoothes the air passing over the wing, reducing turbulence until the bird levels off.

When you've got fancy plumage, you have to be a bit more careful. With its green tail symbolizing lush plant growth, the Resplendent Quetzal of Central America helped inspire the Mesoamerican legend of Quetzalcoatl, the feathered serpent and god of vegetation. The quetzal avoids injuring its iridescent tail—which is nearly twice as long as its body—by launching itself backward to clear its perch and get airborne.

Landing requires even more skill and agility than taking off. Having landed a plane several hundred times, I can attest that putting the plane in the right place at the right speed is the hard part. To land, airplanes and birds have to approach the ground at the correct speed—too fast and they hit the ground too hard (or run out of runway), too slow and they stall and crash. Just prior to landing, a bird moves its body closer to the vertical, extends its alula, and spreads its tail feathers as a brake while beating its wings against the direction of flight. Waterfowl, cranes, herons, and others use their forwardly outstretched feet as well to increase drag. Constant readjustment of wing flapping speed and position usually put the bird down gently. Studies of pigeons landing indicate that they land with more force on a familiar perch than they do on an unfamiliar one. In other words, they are more cautious when landing in a new place. They also need accuracy. A bird has to land in the right spot; a sparrow can't land in a puddle and a loon can't land on the ground.

Landing in a gusty wind is another story. Airplanes have to land in a particular direction—down the runway—no matter what direction the wind might be. Landing in a stiff crosswind is so difficult airplanes have maximum crosswind ratings. Birds have the advantage of almost always being able to land straight into the wind, but they still have to compensate for changes in air currents. Each bird has to spread and twist its wings, tail, and individual flight feathers millisecond by millisecond to put down gradually.

Woodpeckers, nuthatches, creepers, and other birds that feed on tree trunks approach landing differently. They fly close to a tree, spread their wings to slow down, swerve upward along the trunk to dissipate their speed, and grab on when their momentum has reached nearly zero. Woodpeckers have two toes in front and two in back plus a very stiff tail, so they can hang on tightly and creep upward easily. Taking off, they just have to let go, push themselves away from the tree, and spread their wings.

Soaring and Gliding

Paper kites, the first successful implementation of the principles of flight, have been around for at least three millennia. In the years 550–559, the emperor of China's Northern Qi province studied bird flight but had an odd way of experimenting. He strapped bird-shaped kite-wings to the arms of prisoners and pushed the poor guys off cliffs or towers. Many years later and more humanely, men were lifted off the ground by kites attached to ropes. Kites fly tethered by a string, deflect the wind downward, and soar upward. Birds do the same. What keeps them the air? A bird's wing is curved in cross section with the top slightly convex. When the bird is airborne, molecules of air hit the forward edge of the wing and separate, some going above and some below. To meet at the back of the wing, the molecules of air above move

A Bald Eagle landing at its nest.

farther and thus faster than the ones below and as a result the air molecules above the wing are spread farther apart. This difference makes the air above less dense than that below, so air pressure under the wing is greater and the wing is lifted. The faster the wing moves, the greater the lift. The tail helps in the same way.

Moving in three dimensions requires control in all three. Think about riding in a car when it brakes quickly. The front of the vehicle goes down and the back up, demonstrating pitch. Now the car quickly turns a corner and the car tips in the direction of the turn; this is rolling, the tipping of the body to one side with the axis of the roll through the body from head to tail. A car skidding on snow or ice on the road demonstrates yaw, the horizontal movement around the center of the body, like a weather vane moving in the wind. Pitch, roll, and yaw are minor in a car, but in the air are significant movements that a bird must control in order to be a successful aviator.

Once a bird gets airborne, it can glide, ultimately to the ground, or it can soar, staying aloft using air currents. Birds may use thermals, updrafts of warm air that can form when pockets of adjacent air are different temperatures—over mountains, next to bodies of water, or above paved roads. The birds are lifted by the warm air and move in the desired direction by trading altitude for distance. Birds also use obstruction currents to gain altitude. When wind hits a mountain, cliff, or building, soaring birds use the rising air along the edge of the obstruction. There's also just plain wind. Birds can hover almost motionless in a stiff wind, adjusting their wings, and gain considerable altitude that they can trade for distance. Watch vultures and large hawks ply their skills the next time you see them.

Perhaps the most ingenious use of minimal power for flight is dynamic soaring, employed by some seabirds, especially albatrosses. When the wind is about 10 mph over the ocean, an albatross heads into the wind, gains

Soaring Andean Condor showing wing slots between primary feathers.

altitude, and then turns in the direction it needs to go, trading height for distance. Gottfried Sachs of the Technical University of Munich and colleagues using miniaturized GPS transmitters tracked free-flying albatrosses and found that the birds could fly for thousands of miles using dynamic soaring with little energy expenditure. The researchers speculate that robotic airplanes could be invented that could roam the ocean with little fuel.

Soaring is such a successful technique because slots at or near the end of the wings created by the horizontal and vertical separation of the primary feathers act like winglets on modern airplanes and smooth air over the wingtips. But soaring and gliding are limited to those birds with low wing loadings.

FLOCKS AND FLYING IN FORMATION

Augury, from late Middle English, means foretelling a future event by studying bird flocks. During the Roman Empire, a member of the College of Augurs was always consulted before a battle, election, or starting a major building project to assure that the gods were in favor of the plan. A flock of birds was usually taken as a portent, good or bad.

We have all seen masses of shorebirds, blackbirds, or starlings swirling and swarming every which way in seeming unison. Look at a video of the winter roosts of millions of blackbirds in the Great Dismal Swamp in Virginia. How do they control their movements? Their flight seems almost choreographed. High-speed photography and computer simulations of bird flocks tell us that a bird tracks the movement of its neighbors and adjusts its flight and spacing to stay in sync. A lot of jockeying and readjusting goes on but it happens too quickly for us to notice (recall flicker fusion). George Young and colleagues at Princeton University performed computational studies on the movements in starling flocks and concluded that the birds coordinate their movements with their nearest seven neighbors and that the shape, not size, of the flock matters. According to the researchers, the optimum flock shape is that of a thick pancake: too thin and the birds have a limited view of the flock's movements, too thick and the birds can't keep track of the additional neighbors.

Why do birds flock like this in the first place? One reason is safety in numbers: more eyes to spot predators and perhaps confuse or even mob them. More eyes also mean a greater likelihood of locating food or suitable habitat on a migratory route. Additionally, many young songbirds need the social stimuli of the flock to learn the appropriate behavior of their species. Sometimes surviving requires depending on others.

A flock is organized in a hierarchy with higher-ranked birds toward the front. A study involving a 15-bird pigeon flock fitted with geotransmitters provided the exact location of each member of the flock five times per second. It turns out that the leading bird makes the navigational decisions for the flock, but when the leading bird drops back, another high-ranking pigeon takes its place and leads the flock. So the flock has a leader but the leadership position changes often. If you watch a flock of geese you will notice the same phenomenon. The lead position changes often probably because the leader receives no aerodynamic benefit and needs a rest. I remember seeing a television show when I was a kid, when the early nature programs were from the perspective of hunters, showing Canada Geese flying in formation. The narrator said, with all seriousness, that if the leader of the flock was shot, the entire flock would become disoriented and lost. Nope.

Drafting is a common energy-saving technique in bicycle road racing in which a fast biker produces a turbulent wake, thereby producing a low-pressure center that the following biker can ride into and that will help propel him or her forward. Likewise, a soaring bird producing lift creates an upwash of air behind the tip of its wings. These tip vortices, similar to small tornadoes, are helpful to birds flying in a V formation. Theoretical calculations indicate that 25 birds flying in a V, with each bird except the lead flying over the upwash of the bird in front, have a 70 percent increase in range as compared to a lone bird. Henri Weimerskirch and colleagues at the French National Center for Scientific Research implanted heart rate monitors under the feathers of eight pelicans trained to fly in a V over the Senegal River. By measuring the heart rates, they could determine that optimal spacing saves energy and improves communication. A 2014 article in *Scientific American* reports a study of captive-bred Bald Ibises flying with attached GPS loggers that counted wingbeats and found that the birds not only place themselves in the best position to take advantage

V flight of Canada Geese.

of the V formation but also flap in coordination with the wingbeat of the bird in front. Seems like they have it down to a science.

Brown Pelicans, cormorants, and other waterfowl often fly in a line just above the surface of the water. One reason for this behavior is that wind speeds diminish close to the water (or ground) because of friction. Flying close to the surface of the water, the birds also take advantage of "ground effect." When a bird or airplane is less than a wing's length above the surface of the ground or water, tip vortices cannot form and the air is compressed between the wing and the water (or ground), increasing lift. All these techniques save energy, which means less time is needed for foraging, and exposure to the elements and predators is decreased. Smaller birds like pigeons and starlings don't take advantage of V formations or ground effect because their wings are too small to produce a significant upwash of air or benefit from increased pressure from below the wings.

FLYING UNDERWATER AND FLIGHTLESSNESS

Variations on the theme of flight continue with birds that both swim and fly and birds that have completely traded flying in the air for flying underwater.

Each of these lifestyles matches the demands of the birds' environment.

Most birds that are good swimmers are not particularly good fliers because they traded some flight adaptations for swimming ones. However, some birds are both good swimmers and competent fliers such as the ocean-living Pelagic Cormorant and Thick-billed Murre. A study found that the cormorants used a lot more energy to swim than similarly sized penguins did. The murres proved to be more efficient swimmers than the cormorants but still used 30 percent more energy than penguins. To be any more efficient at swimming the murres would have to reduce their wing size or put on more muscle, either one of which would make it impossible for them to fly. The same principle applies to seabirds like petrels, guillemots, and shearwaters: the better the swimmer the worse the flier. Diving petrels are an exception as they can fly quickly over the water and dive in, fold their wings slightly, and swim rapidly under the sea surface for short distances.

Terns, gannets, and some pelicans dive after their prey and are reasonably agile underwater but their hollow bones and buoyant bodies let them penetrate only a short distance under the surface so they must nose-dive into the water, sometimes from considerable heights. I once visited a Northern Gannet colony at Cape St. Mary's, Newfoundland, where 6000 pairs of these

Thick-billed Murres are very efficient swimmers and capable fliers.

birds breed. ("Cape St. Mary's pays for it all" was a saying of fishermen who could catch a boatload of fish off this productive coast in years past). Gannets dive from up to 100 feet high, folding their wings back and piercing the water like an arrow. Their eyes are placed forward for wide binocular vision, they have no external nostrils, and their upper chest cavity has air sacs to cushion impact with the water. They plunge into the water and swim after fish using their webbed feet for propulsion and partially extended wings for steering. After a few moments they bob to the surface.

Most birds that swim in fresh water propel themselves with their feet. Loons are expert swimmers and divers and have been recorded at depths of 600 feet. Their bones are solid, and their legs are laterally flattened, far back on their body, and terminate in webbed feet. Agile in the water but awkward on land, the loon got its name from the Norwegian word "lom" meaning a clumsy or dull-witted person.

Grebes have lobed toes, not webbed feet, but they are still fine swimmers and can dive down to 130 feet. Johannsen and Norberg of Harvard University filmed swimming Great Crested Grebes and noted that their asymmetric toe lobes and separation between them increased lift as each toe acts as a separate hydrofoil. Like loons, grebes have legs that are placed far back on the body, but unlike loons, grebes can run a short distance on land and even on water. Western and Clark's Grebes are well known for their "rushing" display during courtship. A pair will lunge out of the water and run together across the water's surface at 15–20 steps per second for up to 65 feet, without using their wings. I watched this display for several summers on Eagle Lake in northeastern California where I conducted a census of their population, one of the largest in North America. It is always an amazing sight to see birds running upright across the water's surface.

Dippers are unusual in being highly aquatic songbirds. They dive into rushing streams and swim with their strong wings as they search for aquatic insects. A flap covers their nostrils, their plumage is unusually thick, and their blood holds a high concentration of oxygen. Their feet play little role in locomotion, but the long toes and claws help the bird hold onto rocks. An enlarged preen gland ensures waterproof feathers and their nictitating membrane protects the eyes and assists in underwater vision. Once called the water ouzel, it was renamed because of its habit of dipping and bobbing along stream edges. I have seen dippers nesting on ledges behind waterfalls, which required the parents to fly repeatedly through the waterfall to forage downstream and fly back again through the falling water to the nest.

The Atitlan and Junin Grebes of Peru and Bolivia and the flightless cormorant of the Galapagos traded their ability to fly to become more aquatic. Penguins, which spend about three-quarters of their lives at sea, also gave up flying in the air, but not in the water. They move through the sea with their flat, solid, wings and strong flippers that twist at a 20-degree angle, producing thrust on both the up and down strokes, a bit like hummingbirds. Penguins swim with their heads against their shoulders and feet held against the tail to reduce drag and help in steering. Considered by many to be the most hydrodynamic shape in the animal kingdom, the penguin body has inspired the shape of submarines, torpedoes, and underwater vehicles. According to *The Penguins*, the birds generally travel at about 5 mph but can exceed 15 mph.

Only about 0.5 percent of birds are unable to take to the skies, but for some it has worked well: besides penguins, the birds include kiwis, Ostriches, cassowaries, rheas, Emus, tinamous, five waterfowl, Kakapo parrots, one cormorant species, two grebe species, and 21 species of rails, coots, moorhens, and

The flightless Titicaca Grebe, an endangered species.

crakes. Flightlessness did not evolve as a separate branch of the bird world; flightless birds from various families regressed from their flying ancestors. Regression is not unusual in evolution; consider hairless humans, blind cave fish, and legless snakes. We know that natural selection chooses the more advantageous adaptations over the less useful ones, so why did flightlessness confer a survival advantage over flight in the case of these birds?

Several flightless birds once inhabited Hawaii, New Zealand, and other oceanic islands. Before the arrival of the Polynesians in 900 AD there were perhaps 30 species of flightless birds on New Zealand, 25 percent of all the bird species there. New Caledonia, Madagascar, Jamaica, and many other islands had flightless birds, the best known being the Dodo of Mauritius. These birds could not have come from one branch of the avian tree, so flightlessness clearly evolved in different groups of birds in different places.

Flight allows escape from predation by land animals, so in the absence of predation the selection pressure to maintain flight is absent. But since many island birds continued to fly, the absence of predators can't be the whole story. It may be that as more birds arrived on the islands and as competition for food increased, some birds reduced their pectoral muscle mass, reducing their basal metabolic rate. As smaller, flightless birds, they need less energy and thus reduce competition for food.

Flight does not belong to birds alone but this characteristic transcends all other features of birds. The grace, elegance, and apparent effortlessness of a flying bird is perhaps the most moving sight in nature. Memorable documentaries like *Winged Migration*, *Earthflight*, *Fly Away Home*, and *In-Flight Movie* give us views of birds we could never experience otherwise and provide a glimpse of what it might be like to be airborne and untethered. Watch the next birds you see and your appreciation of their freedom from the bounds of the earth will be heightened by this spectacle of evolutionary design that enhanced their lifestyles and ensured their survival.

TRAVEL HITHER AND YON

Migration and Navigation in an Endless Sky

Bird migration is the one truly unifying natural phenomenon in the world, stitching the continents together in a way that even the great weather systems, which roar out from the poles but fizzle at the equator, fail to do. It is an enormously complex subject, perhaps the most compelling drama in all of natural history.

—SCOTT WEIDENSAUL, *Living on the Wind*

Migration is perhaps the most death-defying behavior that birds participate in and it almost seems ironic that the behavior evolved to ensure survival. Flying for days, sometimes nonstop, while facing wind and weather, unfamiliar territory, and predators en route is a big gamble, but apparently a worthwhile one because about 40 percent of the world's avian species participate in annual migrations. Most birds fly at about 15–50 mph during migration and travel 100–500 miles in a day. Larger birds tend to fly faster than smaller ones; ducks and geese average 40–50 mph while flycatchers average more like 17 mph. So the voyage is generally not quick. The Arctic Tern migrates from the Arctic to the Antarctic and back every year. Bar-tailed Godwits fly nonstop from Alaska to New Zealand. And Bar-headed Geese fly over the Himalayas, even above Mount Everest.

A mutation in the gene that controls hemoglobin allows Bar-headed Geese to carry a greater concentration of oxygen than other geese.

Although birds are the quintessential wanderers, many animals migrate to sustain their lives. The massive annual migration of wildebeest, along with enormous herds of zebras, gazelles, elands, and impalas, is closely intertwined with the rainfall patterns of the Serengeti. Great White Sharks feed off Alaska's coast in the summer and return to the coast of California during the colder times of the year. Monarch butterflies migrate from the Rocky Mountains to Pacific Grove, California, taking four generations to complete the trip. In the Shawnee National Forest of southern Illinois, snakes, turtles, and frogs move from the dry limestone cliffs where they spend the winter to breed in the nearby LaRue Swamp in the spring.

Establishing the evolution of a behavior is challenging but we can make some guesses, based on the distribution of fossils and present-day behaviors, as to why migration has proven more advantageous for long-term survival than being sedentary (non-migratory). Here is one scenario: a sedentary population of birds grows and eventually fills a habitat, competing ever more strongly for resources such as food and nest sites. To insure their own survival, some of the younger birds move to another area and establish their own population. If the new area is farther north, it might become unsuitable in the winter, so these northern birds return south to the sedentary population, along with their young. This movement lessens competition for resources for the entire population during the breeding season although competition during the winter season may be more intense. The following year, the population of migratory individuals moves northward again, perhaps occupying an even larger area. Over the years the sedentary population, being invaded in the winter by an ever-larger migratory population, eventually succumbs to competition from the migrants, resulting in an entirely migratory population. We once assumed that today's migratory bird populations all evolved from sedentary populations, implying that migration is an evolutionary advancement to enhance survival. But it appears that patterns of avian migration developed almost randomly and that the migratory habit is elastic. A study published in 2012 of the wood-warbler family of North and South America reveals that the ancestral wood-warbler was migratory and that the loss of the migratory habit—becoming sedentary again—was as frequent as becoming migratory, another case of regressive evolution like flightlessness.

Although we know significantly more now than we did in 1703—when a pamphlet written "By a Person of Learning and Piety" stated that migratory birds flew to the moon for the winter—the whys and hows of migratory behavior are still being discovered. Why do some migrants fly nonstop for days and others seemingly take their time? Is the path of migration instinctive or learned? Do birds follow established routes? Why do some but not all birds migrate? Why do some species migrate but similar species do not? Why do some individuals of a species migrate and others not, and why is it that one individual may be migratory one year and sedentary the next? Both genetic and environmental components are at play here, but their roles have yet to be sorted out. Until scientists found ways to track birds, these were all mysteries, and all the answers are not in yet.

HOW WE LEARNED ABOUT MIGRATION: BANDING BIRDS

The study of migration began by marking birds with tags or bands. At least as far back as 1000 BC the Chinese kept birds for falconry, each one marked to indicate ownership. Early Romans sent simple messages across the countryside by tying threads to the feet of crows. Perhaps the earliest record of a metal band on a bird's leg was in 1595 when a Peregrine Falcon belonging to Henry IV of France disappeared in pursuit of a bustard (a large, swift-running bird of the Old World plains) and was later found on the island of Malta, south of Italy, 1350 miles away. A duck shot in Sussex in the United Kingdom in 1709 had a silver band around its neck inscribed with the arms of the King of Denmark. J. J. Audubon tied silver cords around the legs of young Eastern Phoebes and later identified them when they were adults. But the first attempt to gather any real information about bird movement might have been the marking of a Great Gray Heron captured in Germany in the early 18th century. It had metal bands on its legs, one of which indicated it was banded in Turkey a few years earlier. In 1899 Denmark, C. C. Mortenson was apparently the first person to systematically mark birds with numbered bands to collect data on their travels.

In 1920 the Federal Fish and Wildlife Services of Canada and the United States began to oversee bird-banding activities and the collection of data in the two countries. Since 1960, approximately 64 million birds have been banded, and either through field observation or recovery of bands, there have been approximately 681,000 encounters with banded birds. From these encounters we can determine information about migratory routes, longevity, social structure, population sizes, and disease. The Bird Banding Laboratory of the U.S. Geological Survey collects comprehensive data, requiring painstaking attention to detail on the part of licensed banders, resulting in extensive and accurate information on bird movements. In addition to the date of banding and recapture, information like age, sex, stage of molting, whether blood or feather samples were taken, if auxiliary leg or wing or neck markers will be used, and signs of disease are all noted.

I was a licensed bird-bander for several years and banded a few hundred songbirds. Although I did recapture a few birds in the same location a year or two later, I never got a recovery from outside my banding site. This is typical: band returns from songbirds are around 1 percent because the birds have relatively short life spans and small bands are hard to find after a bird has died. However, about 15–20 percent of waterfowl bands are recovered each year because the birds are hunted and if a duck dies in the wild, the aluminum band is big enough and may last long enough to be found someday.

Yellow-throated Fulvetta of Southeast Asia showing metal leg band.

Banding, or ringing as it is called in Europe, has supplied important information on the movements of birds but little about what the birds did between banding and later band recovery. Research with new techniques and equipment—including radar, chemical isotopes, radio frequency identification tags, very high frequency radios, videography, and more recently GPS loggers and satellite tracking—has provided considerably more information. The British Antarctic Survey uses fingertip-sized data loggers attached to the legs of seabirds to record light intensities at different latitudes and longitudes to provide position information. Even smaller data loggers, as light as 0.05 ounces, are being made for use on songbirds. And sophisticated technology using isotopic ratios adds even more information. A study of Black-throated Blue Warblers measured naturally occurring isotopes of carbon and hydrogen in their feathers. The ratios of the isotopes reflected the diets of the birds and determined that the northern breeding populations winter farther west in the Caribbean than the southern breeding populations. Most of these techniques are expensive and require considerable training on the part of users, so bird banding will continue to be a primary source of data on bird movements for a while.

Banding is allowed under the provisions of the Migratory Bird Treaty Act but one needs to have a legitimate reason for banding birds; it is not looked upon as a casual hobby. Bird banding can have a negative impact on the survival of birds as bands can cause injury from friction against the leg and add weight to the bird, and the process of capturing the birds causes stress. Neck bands on Trumpeter Swans have iced up, causing severe mobility problems, and aluminum bands put on the flippers of King Penguins appeared to result in a 16 percent higher mortality rate. Banders argue that these problems are exaggerated and that the information gleaned from banding is worth the risk, but I suspect banding will slowly be supplanted by new techniques as they become more accessible and less expensive.

ALL IN THE TIMING: HOW DO BIRDS KNOW WHEN TO MIGRATE?

Are you a sports fan? Here's something to consider: statisticians studied 40 years of football scores and determined that West Coast teams have an advantage over East Coast teams when the game is played after 8 P.M. on the East Coast of the United States. Similar results were found with baseball scores. What's happening? It's the players' internal clock. West Coast teams are more alert in the evening on the East Coast than are East Coast teams. Did you get up at about the same time this morning as you did yesterday? Do you feel tired in the mid-afternoon? Do you go to bed at about the same time every evening? Probably. You are showing circadian (Latin for around the day) rhythms that are reflected in your brain activity, body temperature, hormone levels, blood pressure, and other metabolic functions. This is your body's internal clock. These patterns of daily activity are tied to the photoperiod (length of daylight). You interrupt your circadian cycle every time you experience jet lag by flying several time zones away from home, but in a few days you adjust to the new daylight regime. (The quickest way to adjust is to adapt to the local time regime and expose yourself to sunlight, as the sun will reset your cycle.)

Now think about *why* you awoke and got out of bed this morning. The proximate factor is your daily rhythm: you usually get up at that time, perhaps assisted by an alarm clock or your significant other bringing you breakfast in bed. As unlikely as the latter might be, these are the immediate reasons you rousted yourself out of the sheets. But why get up at all? Because you have things to do—your boss expects you at work, you have a fishing trip planned, you have to get the kids to school, you are hungry. These are ultimate factors—the actual reason(s) you get up and out of bed.

The proximate factors of migration are those environmental cues that set birds off on their journey, primarily day length; and the ultimate factors—the reasons they leave on this annual journey—are such things as food and a milder climate. The timing of migration is pretty predictable. Birds arrive on their breeding grounds at about the same time every year, the key word here being "about." Some species show little variation in average arrival dates and other species demonstrate considerable variability. Insectivorous birds appear to be more predictable than other kinds of foragers. Tradition has it that swallows return to California's Mission San Juan Capistrano on March 19th, but the actual times vary, often by several days. (Unfortunately, remodeling and increasing development in the area has caused the Cliff Swallow

Ad for buzzard festival in Hinckley, Ohio.

population to decline to virtually zero, but efforts are being made to entice the birds back.) Since 1957 the town of Hinckley, Ohio, has been holding a "Return of the Buzzards" festival to celebrate the arrival of Turkey Vultures, predicted to be March 15th each year. Vagaries of the weather affect the travel of migrants so the dates of arrival are not exact, but they are pretty close. Photoperiod is the main proximate factor.

Photoperiod or Day Length

The length of daylight each day changes gradually all over the earth except at the equator and poles. At the equator, the day length is constant at 12 hours, 7 minutes. At the North Pole there are 163 days of total darkness and 187 days of "midnight sun." Tromso, Norway, experiences "polar nights" in which the sun does not rise for 60 days in the winter. Day length in London in mid-August is 14 hours, 51 minutes versus 7 hours, 51 minutes in mid-December. In Mexico City the corresponding figures are 12 hours, 9 minutes and 10 hours, 59 minutes. These changes in photoperiods can have significant effects on organisms.

We have known for a long time the importance of circadian rhythms in humans and other animals. Irregular changes in light and dark periods, such as some workers experience with frequent shift changes, can interrupt the growth of blood vessels and lead to diseases or conditions such as heart attack, stroke, and the delayed healing of wounds. The importance of circannual cycles has also been demonstrated, such as differences in the rate of cell division in the bone marrow and intestine, the volume of red blood cells, blood pressure, and cholesterol levels over a year's time. (Don't confuse these annual cycles with biorhythms, the wacky idea that our lives go though certain cycles based upon the day of our birth.)

Unlike weather, day length is predictable. The times of dawn and dusk change the same way at the same time in the same place. Birds are cued by the lengthening photoperiod to prepare for and depart on their migratory journey. This cue is called the *Zeitgeber* (German for time-giver). With increasing photoperiod, birds begin to exhibit physiological and behavioral changes. In the equatorial area with little change in the photoperiod, the cue may be the intensity of the light in the wet and dry seasons.

As the time to leave approaches, birds change their behavior. They begin to eat more to put on weight, particularly fat, for energy. For many years researchers observed captive birds as migration time drew near and noticed that the birds began to show restlessness, known as *Zugunruhe* (German for movement anxiety). The activity and orientation of caged birds during this time were measured by various electronic and mechanical means. In 1966 the Emlen team of father-and-son researchers invented the Emlen funnel, an elegant but simple mechanism to measure *Zugunruhe*. This funnel-cage was lined with paper on the sides and sloped down to an ink pad on the bottom. Whenever a bird jumped up the side of the funnel, its inked feet marked the paper. When the paper was later removed, the relative density of ink markings on various areas of the funnel indicated the bird's directional preferences. The photoperiod, orientation of daylight or moonlight, and the night sky could be manipulated (via a planetarium) to see what the effects would be on the captive birds' orientation.

As the spring photoperiod lengthens, birds begin their northward exodus. Although weather is unpredictable, the date(s) of departure is, on average,

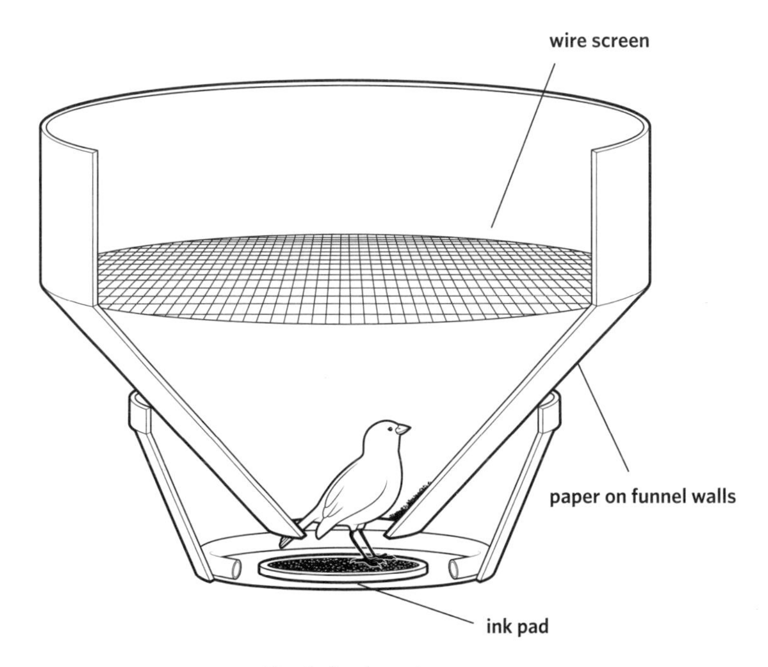

The Emlen funnel.

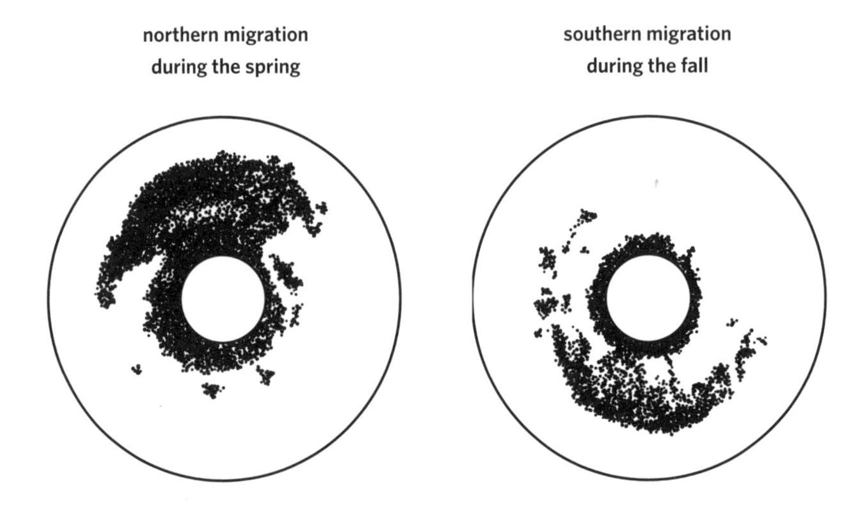

Density of ink markings indicate the bird's directional preferences.

the best date(s) to go. If the birds were to leave too early because of a bout of mild weather, they could encounter bad weather during the passage. If they delayed their departure significantly during a low-pressure spell, they might find the resources depleted when they arrive. Insects emerge and become active, and plants germinate, emerge, and bloom as the weather warms and the birds need to be there to take advantage of the food supply. Extenuating factors such as soil moisture and snow cover as well as temperature affect the appearance of plants and insects, but they emerge at about the same time each year. Thus the timing of when birds migrate each spring is tied to the photoperiod and their food supply to the increasing temperature. As with most genetic characteristics, there is variation among the individuals of a species, including their date of departure. In cold years the earliest arrivals may be at a disadvantage; in warm weather tardy individuals may find a depleted food supply. In years of weather extremes, either early- or late-arriving individuals of any one species may have a higher (or lower) survival rate.

The entire population of many bird species moves from the wintering to breeding grounds in about three to six weeks. The White Wagtail arrives in Finland from North Africa over a period of about three weeks, similar to that of the Barn Swallow that winters in the southern half of Africa and arrives in Europe a little later. The Common Swift, wintering in the southern third of Africa, arrives later still, but with only a two-week difference between the arrival of the first and last swift. All three species depend upon insects for food, but the White Wagtail is a bit more catholic in its tastes and will eat some vegetable matter; the Barn Swallow feeds mainly in mid-air but will pick insects off of vegetation, rocks, and the water's surface; the Common Swift is restricted to aerial foraging, a dietary difference that probably explains the difference in arrival times to coincide with the peak of insect emergence.

UNDERSTANDING MIGRATORY BEHAVIOR

Migratory behavior varies widely, depending on the geography and the bird species. There are long- and short-distance migrants, species with both migratory and sedentary populations, summer-winter and wet-dry seasonal migrants, and altitudinal migrants. All the individuals of a population might migrate at about the same time or spread out the timing depending on age and/or sex. Red-winged Blackbird males, for example, winter in all-male flocks and travel north early to set up breeding territories to attract females upon their later arrival. I could go on, but you get the idea—migration is very labile and changes as necessary to assure the survival of a particular population.

In the midst of writing this chapter during the month of October, I traveled to Bolivia and spent a week on one of the Amazon's upper tributaries. While fishing and birdwatching, I was fascinated by the variety of both kinds of animals, each taxon totaling around 1400 species. I didn't see all those birds or catch that many different fish, but I did see an amazing number of both. Interspersed with the native Bolivian birds were migrants from North America—the Peregrine Falcon, Osprey, Lesser and Greater Sandpipers, Solitary Sandpiper, and Barn and Bank Swallows. It has always fascinated me that these interlopers from the north seem to adroitly slip into the tropical Amazonian ecosystem unnoticed and unbothered by the residents. This certainly says something about the productivity of the tropical forest.

Migratory behavior varies considerably among species and between populations within a species. Song Sparrows and Blue Jays in North America are sedentary while orioles and tanagers migrate, and in the United Kingdom the Chough and kingfisher are sedentary while swallows and wagtails move southward. American Robins in North America are non-migratory in the southern United States but breeding populations of Alaska and Canada move southward in the fall. Chiffchaff generally breed in northern Europe and Asia and winter in southern Asia and Africa, but in recent years some have been overwintering in the United Kingdom. Black-headed Grosbeaks breed across western North America and winter in Mexico, meeting up with sedentary populations of their species.

The Southern Hemisphere has a smaller land area than the Northern Hemisphere, and migration patterns are not as strongly developed. No birds breeding in the Southern Hemisphere winter on a different continent and some Southern Hemisphere migrants never cross the equator. Rainfall also has a greater influence on migratory behavior in the Southern Hemisphere because insect populations increase with more rain. The Pied or Jacobin Cuckoo arrives in India from Africa at the beginning of the monsoon season, and the bird and the weather phenomenon have become inextricable in Indian folklore as the harbinger of the monsoon. Nightjars, birds named for their crepuscular feeding habits and the "jarring" sound of their voices, are good examples. The Standard-wing Nightjar breeds from Senegal east to Ethiopia and migrates northward to avoid the wet season. The Pennant-winged Nightjar breeds south of the African equator and winters somewhat north of it; the bird's migratory journeys are very protracted and may or may not be related to the wet season. In Australia, only 7 percent of migratory species have totally separate breeding and wintering areas and, on average, they move only about 9 degrees of latitude, compared to Northern Hemisphere migrants, which average 22 degrees of latitudinal movement.

Altitudinal migrants, such as the Red Grouse, Spotted Owl, and Steller's Jay, move from a higher elevation in the summer to a lower one for the winter. The White-throated Dipper of Europe and the American Dipper migrate slightly southward or downslope from a higher elevation, depending on the severity of the winter. But the American Dipper's situation is more complicated and gives us some idea of how partial migration might have evolved. Studies indicate that although a large population of dippers might be resident in the lower areas of a watershed—a main river or stream, for example—when

John Gould's (English ornithologist and painter) depiction of the White-throated Dipper.

breeding season arrives some birds move to higher elevations because of competition for food and nest sites. Even though nesting success is comparable, riverside breeding birds at the lower elevations often have second broods. So the partial migration upslope benefits the entire population. In the tropics up to 20 percent of the bird species participate in altitudinal migrations. One study indicates that the elevation at which tropical birds in Costa Rica breed is inversely proportional to predation on their nests, so higher nests have less predation. But higher elevations may also be subject to the vagaries of the weather as evidence also shows that an unusually rainy season will cause birds to migrate downslope from higher elevations.

How Long Does Migration Take?

Birds begin and end their various journeys at different locations, so the time it takes to travel varies, and their speed depends on environmental conditions. Weather can slow or accelerate the trip considerably. In early February, Canada Geese leave their wintering grounds in the southern United States and arrive in northern Canada at the end of April. They more or less follow the 35°F isotherm (a line of equal temperature at a given time or date on a weather map) northward. This makes sense because if they were to fly in front of the isotherm, they would confront freezing conditions and a dearth of food. The Baltimore Oriole spends summers in eastern North America and winters in the Caribbean, Central America, and northern South America. At about 20 mph, migrating mainly at night, they might cover 150 miles a day, taking two to three weeks for the entire trip.

Shorebirds on both sides of the Atlantic make incredibly long migrations from the Arctic tundra to the Southern Hemisphere. Bar-tailed Godwits nest in the Alaskan tundra and winter in New Zealand. In the spring they leave New Zealand and fly northwestward, stopping to refuel in Australia and the Korean and Russian peninsulas before arriving about four months later in Alaska. The return winter journey takes them directly from Alaska to New Zealand during the month of September—a nonstop journey of 6600 miles. In the summer of 2006 researchers attached satellite transmitters to 16 Bar-tailed Godwits. On the trip from Alaska to New Zealand, one female bird was recorded as traveling 7145 miles without stopping, flying at an average speed of 34.8 mph for nearly eight days.

Bar-tailed Godwit in winter plumage.

The Arctic Tern holds the record for the longest migratory route. Nothing is spectacular about the appearance of this medium-sized bird, but its lifestyle in the Arctic and Subarctic and wintering in the Antarctic have made it famous. They are circumpolar breeders in the Arctic from May to August where summer days are long. The birds typically lay two eggs and defend their nest vigorously. Although they feed mainly on fish and crustaceans from the ocean, they will consume insects during the nesting season. After the young are fledged, the colonies leave the breeding area to winter in the Antarctic region from November to March when the photoperiod is long there. From about April to May the terns migrate northward again, traveling an average of about 300 miles per day, although individual birds have been observed to fly as much as 400 miles per day. Some individually tracked birds flew up to 66,000 miles each year. As they follow the seasons and photoperiodic changes, Arctic Terns see more sunlight than any other animal.

Arctic Terns can live for more than 30 years, and it has been said that during their lifetimes some of these birds may travel the equivalent of three round trips to the moon! An impressive accomplishment, but how many of them actually live to be 30 years old? I found an old banding study that estimated the annual mortality rate of Arctic Tern adults at 18 percent. With

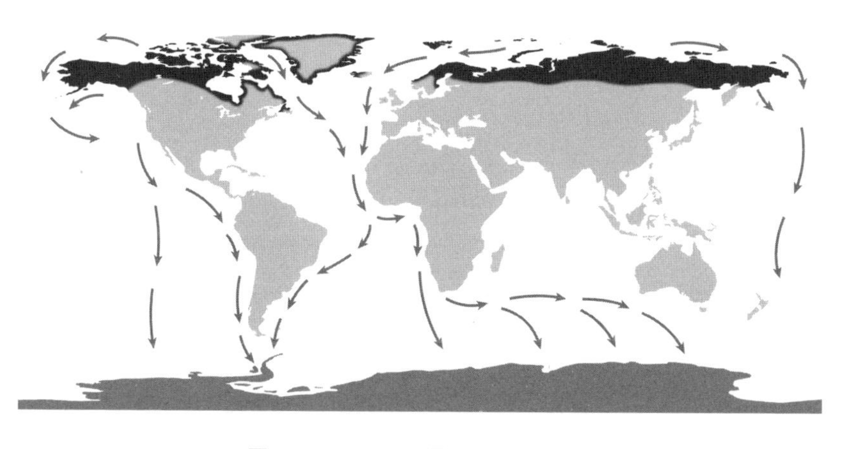

Arctic Tern migratory routes showing only southward migration.

that death rate, if we start with a population of 1000 birds in the Arctic and 18 percent of this group dies each year, 30 years later perhaps one bird will have survived long enough to have made 30 round trips; what an amazing life that bird led.

Maximizing Mileage

An endless amount of information has been amassed on the locations of breeding and wintering grounds, the routes to and from, the speed of travel, and the dates of departure and return. But the big question has always been, "How do birds get the energy to cross large bodies of water and fly nonstop for days?"

Migrating birds face three challenges: the energy cost of flight, the need to refuel en route, and the endurance needed to complete the journey. The importance of fat in fueling bird, insect, and mammal migration cannot be overstated. Fat is an excellent energy source; compared to protein and carbohydrates, fats provide more than twice the energy per unit of weight, and because fat does not absorb water, supplies of fat weigh less than other fuels per calorie. But birds also need to accumulate protein to replace the loss of muscle mass during migration. The ratio of fat to protein accumulated before migration varies from 1:1 to 10:1, depending on the diet of the birds, their route, and distance of flight.

Fat accumulation is the main source of weight gain before migration. A rather extreme example of this is the Eskimo Curlew, an abundant shorebird in North America 200 years ago. J. J. Audubon and Elliot Coues, 19th-century naturalists, described enormous expanses of midwestern plains covered with curlews resting while en route north from South America to the tundra of western Canada and Alaska. When the birds were making their return trip, market hunters would kill two million curlews a year for food. The Eskimo Curlew was nicknamed the "doughbird" because it accumulated so much fat before its southward migration that when it was shot and plopped to the ground, the heavy layer of fat resembled dough. The hunters probably exterminated it, as it has not been reliably reported since 1963.

Birds store fat under their skin, as you probably have noticed when preparing a whole chicken. At first the fat accumulates in particular areas, such as in the depression formed by the furcula, and then in the skin below the breast muscles; additional fat is stored all around the skin. Measuring the amount of fat a bird has is important in understanding the migratory strategies of different species, so ornithologists have devised a crude method of assessing fat stores of captured birds by looking at the furcular depression through the translucent skin. All one needs to do is blow on the bird's throat to separate the feathers and expose the skin. A score of one means no fat, the furcular depression full of fat is a three, and a score of five signifies a full load of blubber.

Garden Warblers of the Palearctic, weighing about 0.6 ounces, put on fat and increase their body weight 30–40 percent before initiating migration to their wintering grounds in Africa south of the Sahara. They need to fuel up as much as possible before crossing the Mediterranean Sea and the sands of the Sahara where they have little or no chance to feed; upon arrival they will have lost virtually all their body fat. On their spring return, the birds feed heavily before crossing the desert, some individuals increasing their weight by 50 percent and 10 percent of them doubling their weight. (Imagine if you weighed 130 pounds and you increased your weight by 50 percent—how much time would you have to spend at the gym to get back to your original weight?) In addition to fat, their flight muscles increase in size by 15 percent, giving them not only additional flying power, but protein that can be used if fat stores are depleted. Rufous Hummingbirds increase their body weight by 60 percent in preparation for flying from the Rocky Mountains to Mexico, but converting sugar to fat is expensive, costing 16 percent of the calories they ingest.

After birds leave the nesting area and head to their wintering grounds, most stop en route to refuel and rest. Food might be limited as in the case of shorebirds, which often have to wait until the tide goes out. Prior to migration, birds are efficient eating machines with their gut and liver increasing in size to digest food and store fat. As the storage capacity of fat reaches its limit, the heart, blood vessels, and flight muscles increase in size while the gut and liver atrophy as a means of weight reduction and some of their proteins used to build muscle. In addition, enzymes are activated that control the deposition and utilization of fat, and the number of red blood cells increases so that more oxygen can be carried.

Many birds like thrushes, rails, sparrows, flycatchers, orioles, waterfowl, warblers, blackbirds, and shorebirds migrate at night even though they are usually diurnal because, except for aerial insectivores, they have found it adaptive to fuel up during the day and fly at night when the winds have diminished. If smaller migrants flew during the day, stopping to feed at frequent intervals, their journey would take longer, and, if they spent the night resting, they would need to refuel again in the morning before resuming their trip. But some birds, like the tiny hummingbird, may need to enter a reduced state of metabolism, torpor, to make it through their journey.

One factor that affects gas mileage in an automobile is weight: cars that weigh more use more fuel. Another is shape. Birds fill up on fat but the

Anna's Hummingbirds may enter a state of torpor during some migratory nights to preserve energy.

efficient way the fat is distributed on the body minimizes drag. But weighing more means fewer "miles per gallon." So birds can employ one of two survival strategies during migration. One is to spend less time en route by flying longer and farther nonstop to reduce the chances of encountering bad weather or predators—but this choice requires birds to store more fat and expend more energy carrying it. The other approach is to carry a minimal load of fat and stop frequently to refuel, a common pattern among shorebirds, songbirds, and waterfowl. The best tactic for many species seems to be some combination of these two approaches—short hops alternating with extended flights. Birds that feed on the wing like martins, nighthawks, swallows, and swifts, employ a "fly and forage" strategy so they are able to make long flights with minimum fat. Satellite tracking of Ospreys traveling from Sweden to West Africa revealed that the birds flew 60 percent of the time and were stationary for 40 percent. But there was a lot of variability; some birds never stopped. It had to be that they fed along the way, catching fish from lakes and rivers. Again, natural selection honed migratory behavior to be flexible in order to allow the greatest number of individuals to survive the trip.

Flying is energy intensive; the average hummingbird uses 5 to10 times the amount of energy in flight than at rest and its oxygen consumption per unit of body weight is about 10 times that of a human athlete in action. In the spring, Ruby-throated Hummingbirds fly north across the Gulf of Mexico, about 500 miles, nonstop, in about 20–24 hours. For years birdwatchers could not fathom how such small birds could perform such an amazing feat. Again, it's fat stores increasing their weight by 40 percent. Many other bird species take this overwater route and one of the first places they land is Dauphin Island, Alabama, a prime birding spot. One spring I attended a meeting there and saw about 75 exhausted but gorgeous male Rose-breasted Grosbeaks, having just crossed the Gulf, in one tree. These were the days before bird festivals and the locals were bemused by us birdwatchers showing up in bars and restaurants in khaki outfits and befitted with binoculars. We were the butt of local jokes. Today the island is a bird sanctuary and the annual Alabama Coastal Bird Fest brings in tourists and a good deal of money; the locals now find birdwatchers less funny.

Blackbirds resting in a tree.

Migratory routes, distances, and strategies vary among and within species. Like all adaptations that evolution has crafted, the flexibility provided by variation in migratory behaviors allows birds to respond to changing conditions over time. As the climate changes, lakes dry, rivers alter course, and food sources move; a stop-and-rest route today might become a nonstop journey in a decade. Of course, in order to make any passage successful, birds have to find their way.

NAVIGATION: NO COMPASS, NO MAP, NO PROBLEM

"The winds and waves are always on the side of the ablest navigators," says Edward Gibbon in *The Decline and Fall of the Roman Empire*. Most anyone with a bit of backwoods experience can employ a number of clues to orient themselves, even if the compass gets waterlogged and the map eaten by Bigfoot. Directional clues come in the form of the orientation and movement of the sun, moon, and stars across the sky; the flow of a stream; the slope of a mountain; the prevailing wind and clouds; the moss on the north side of trees; even smell. Birds, without any instruments at all, navigate just fine.

During World War I in the First Battle of the Marne in 1914, the French used 70 mobile pigeon lofts from which they dispatched messages rolled into small cylinders attached to the birds' legs. Amazingly, even though the lofts were on the move, 95 percent of the messages arrived. Cher Ami, the most famous pigeon of all, having lost an eye and leg in combat, received the Croix de Guerre medal from the French for delivering a message that saved 200 American soldiers from an artillery bombardment. Cher Ami traveled by ship back to the United States with General Pershing, the American forces commander, and now resides in a glass case at the Smithsonian Institution. Homing, racing, or carrier pigeons are all the same species, *Columba livia*—the Rock Dove—and are known for their ability to navigate. This is why you will read about so many navigation studies of Rock Doves (although they are often just called pigeons).

Scientists generally agree that birds probably utilize a combination of navigational techniques, either independent or closely intertwined. Landmarks, the sun and stars, geomagnetism, olfactory clues, and low frequency sounds are all possible navigation tools. Besides discovering how different birds find their way, we might learn about our own senses. We all know people with a good sense of direction and others who get lost after turning a corner. What we don't yet know is if this ability is innate or learned. Navigation is among the more difficult areas of ornithology to explore, but also one of the most exciting.

Landmarks

We find our way around our neighborhood and city every day by landmarks. Birds do the same. Learning where to locate food, water, protection from predators, and shelter from the rain, cold, and wind is essential for daily survival, so it seems logical that birds would use landmarks to orient themselves. Even when a bird wanders out of a familiar area, it probably takes note of streams, lakes, large rocks and trees, and other key parts of the environment.

Birds released in a familiar area immediately head in the appropriate direction. Released in unfamiliar territory, they circle until they recognize some sort of landmark. The farther the birds are away from their destination, the longer it takes them to find a landmark. In the case of pigeons, 635 miles is a distance they can, with difficulty, navigate in a day or so. Any farther than that and they generally get lost. A 2005 Oxford University experiment fitted 50 pigeons with tiny tracking devices that showed some pigeons following specific highways, roundabouts, and exits to find their way to their loft. Each time they were released at a distant location, they followed the same routes home, even though a totally straight return route would have been faster.

Data from radio-tagged Canada Geese demonstrate that the birds use landmarks during their diurnal flights, and are especially focused as they approach their destination. The birds also compensate for wind drift by correcting their path as they pass familiar sights. In one case, two populations of Canada Geese migrated as one group through Minnesota; at a certain point the populations diverged. At about that point a large tree fell one year and the following year the migratory populations appeared disoriented for a time before they diverged on their own paths; the tree was apparently a significant landmark.

But the use of landmarks cannot explain navigation on the first migratory flight of immature birds that do not follow their parents or those birds that travel over the sea a considerable distance from the shore. Humans, too, have found their way across vast expanses of the earth by the use of landmarks, but when there were no definitive physical waypoints we had to find other ways to learn our location.

The Sun and Stars

For as long as people have traveled the earth, the sky has provided clues. The ancient Phoenicians of the Mediterranean were excellent sailors and navigators. Although they preferred to sail in sight of land, it was sometimes not possible, so they kept track of birds over the sea, which often indicated that land was not far off. Norsemen noted that a bird with a beak full of fish was headed to its nest and young, whereas a bird with an empty beak was headed away from land. If birds were not visible, sailors oriented by the North Star at night and during the day tracked the course of the sun across the sky. The position and movements of the sun, moon, stars, and constellations were always available under a clear sky.

Several experiments and observations have shown that birds use the sun's location in the sky as a directional clue and support an idea called the sun compass. At Germany's Max Planck Institute in the 1950s, Gustav Kramer discovered that caged European Starlings were able to orient in the direction they wanted to migrate—essentially by inputting information into their sun compass—if they could see the sun move during the day. How they account for the different positions of the sun as seasons change is unknown. The ability to use the sun compass is innate but does not show up in pigeons until about three months of age; it has also been found in fishes, salamanders, frogs, toads, turtles, voles, and bats. Knowing pigeons use landmarks, Knut Schmidt-Nielsen of Duke University and William Keeton of Cornell fitted birds with frosted lenses that prevented them from recognizing landmarks more than 20 feet away. When the birds were released the researchers took note of the direction they flew. On a clear day the birds oriented in the

The European Starling has often been used for experiments and observations to unravel the secrets of migration.

direction of the loft; on a cloudy day their orientations were more random. In another experiment a group of birds with frosted lenses was held in an artificial light regime that shifted the day by six hours. When these birds were released, they flew in a direction that was a six-hour sun shift from the loft direction.

Although some invertebrates like amphipods and marine worms respond to moon clues, there is no "moon compass" for birds. The moon rises about an hour earlier each night as it goes through various phases, so it is not a good clue for orientation. Investigators have even found that the moon tends to confuse birds. Mallards using stars for navigation were less accurate in their orientation when the moon was half or more full.

The Big Dipper and Orion's Belt are about the limit of my astronomical knowledge but for millennia travelers have used constellations for navigation. Birds probably do the same; they certainly use the starry night to find their way. An elegant experiment is one that is simple, concise, ingenious, and persuasive, and that's what I would call German ornithologists Franz and Eleanore Sauer's 1957 planetarium-based investigation. The Sauers put warblers in cages in a planetarium and each evening moved the night sky about 300 apparent miles in the direction that the birds were oriented during their *Zugunruhe*. The birds "migrated" their way around the Mediterranean Sea to Africa. This was the first real demonstration of the avian "star compass." Later experiments were performed with birds whose annual cycle was manipulated so that their physiological condition primed them for either a northward or southward migration. Under the planetarium sky, the birds oriented in the direction of their physiologically primed condition. Shifting the birds' internal clock with a different light regime had no effect on their star compass as it had on their sun compass. Further experimentation indicated that birds orient to a pattern of stars, perhaps a constellation or other group of stars, and follow that pattern as the stars move through the sky. A learning component is involved, as birds need to have some education and experience with the movement of the sun and the stars before they can effectively use them to navigate.

Before radar, night migration of birds was studied by aiming a telescope at a full moon and counting the birds that flew across its face. Sometimes

Birds crossing the moon.

the species could be determined by their silhouette and even the altitude of the birds could be calculated. On one night in 1952, 1400 birdwatchers and astronomers at 265 observation points in North America observed 35,400 birds crossing the moon. That's about 133 birds per observation point, not a lot and only broadly indicative of migration patterns, but it did provide a sample of what was happening.

Geomagnetic Forces

Ornithologists have known for years that birds use some sort of magnetic sense to navigate. Franz Mesmer of 18th-century "mesmerism" fame claimed that animals produced a magnetic force that he called "animal magnetism," a term that has morphed into "body energy fields" and other questionable notions. No evidence has shown that weak magnetic fields have any discernible effect on humans. But on birds? Read on.

Alexander Neckam, an English scholar of the late 12th and early 13th centuries, was apparently the first person to demonstrate the magnetized needle that aligns with the earth's magnetic field. Humans have used compasses ever since and today we have evidence that turtles, lobsters, salmon, and birds use geomagnetic lines to navigate. Early geomagnetic studies on birds in the 1940s were ignored, but later investigations with European Robins by Wolfgang and Roswitha Wiltschko of Frankfurt, Germany, demonstrated that the birds could orient properly even in cages that did not allow them to see the sky. When they artificially reversed the magnetic field, the birds oriented in the opposite direction.

In the late 1960s, William Keeton devised a clever series of experiments, including gluing bar magnets to the back of pigeons before releasing them away from their home loft. (He attached non-magnetic brass bars to a control group of birds.) When the sun was shining, all the birds found their way back with no problem. Under overcast skies the birds with magnetic bars became disoriented but the control group did not. Keeton speculated that the magnets were interfering with how the birds orient to the geomagnetic lines of force. Further experiments by Charles Walcott of Cornell University involved putting Helmholtz coils, devices that produce magnetic fields, around the head of pigeons. When the coils were activated on a cloudy day, the direction the birds flew depended on the direction of the magnetic fields around their heads, while on a clear day the coils had no effect.

So birds use geomagnetic lines of force for orientation, but what is the sensory mechanism for detecting them? A number of animals have magnetite, a magnetic iron-containing mineral, in their heads or cranial nerves. Magnetite deposits have been found in the olfactory nerves of trout, as well as in sea turtles, newts, and several bird species. In birds, the magnetite crystals are located along the ophthalmic nerve running on the right side of the upper bill. Studies have shown that pigeons with opaque contact lenses over their right eye have difficulty navigating, but lenses over the left eye have little effect. There do not appear to be any specific sense organs to interpret the magnetic field, so magnetite crystals may sense only the strength of the magnetic field. The orientation of the magnetic field may be sensed by the birds' innate magnetic compass. Cryptochromes, pigments in the ultraviolet-violet cone cells of bird retinas, allow birds to detect the earth's magnetic field, although whether the pigments indicate the direction or strength of the field is unknown.

Olfaction

Do you remember your mother making your favorite lasagna or your grandmother baking pies? Like scenes and sounds, we often remember scents—both good and bad. Every time I smell a skunk I revisit the time when I was traveling in my uncle's car when he made roadkill out of one of those animals.

We now know that birds can use odors in navigation, but this was not seriously discussed until about 1973 when an Italian researcher postulated the "olfactory hypothesis." He said that pigeons create an olfactory map of the area around their loft and on any journey away from the loft the birds use aromas to help guide their return. Experiments in which pigeons were released in an unfamiliar area upwind of their loft disorientated them, whereas birds released downwind were more successful returning home. It is hard to explain how birds can create an olfactory map a long distance from the loft, especially with shifting winds and changing scents, but apparently pigeons can pick up clues from as few as three volatile compounds in the air, most likely the most abundant human-generated compounds in the area.

I saw very few seabirds in mid-ocean on my cruise line adventures. That's because most of the ocean's nutrients are concentrated near the shore where runoff from the continents and the upwelling of these nutrients by surface winds create a productive coastal zone, with blooms of phytoplankton (microscopic floating plants). Gabrielle Nevitt and colleagues at University of California, Davis, showed that when krill, small ocean crustaceans, eat phytoplankton the chemical dimethyl sulfide (DMS) is released. DMS is detected by seabirds, which come to feed on the krill. The concentration of DMS reflects the topography of the ocean, remains at stable concentration for weeks, and shows predictable seasonal patterns. Experiments with prions (small petrels that eat mainly plankton) indicate that they can detect even low concentrations of DMS. This strongly suggests that seabirds find their way over large expanses of ocean by following a chemical map superimposed over the sea.

Infrasound

Infrasound refers to sounds that humans can't hear—sounds below 20 Hz, such as some ocean waves (with an average frequency of 16 Hz), distant storms, and earthquakes. Pigeons can hear frequencies as low as 0.05 Hz. Early experiments used conditioned responses—whenever a low frequency ultrasound was presented, it was followed by a mild electric shock. After training, the pigeons were again subjected to a series of sounds; each time an ultrasound was presented, their heart rate increased. This indicated that the pigeons expected an electric shock and proved that birds can hear ultrasound. But in the wild, birds need a "sound map," a way to associate landmarks with the aural environment of an area.

Microseisms are small earth tremors caused by natural phenomena such as ocean waves. Since ocean waves are nonstop, they produce a continuous hum all over the earth. It's not likely that birds follow the infrasonic hum because it is spread so widely; however, it is possible that topographic features like mountains, river valleys, and large buildings affect the hum so that a microseism map is produced. Perhaps birds follow the echoes landmarks make, as well as their visual manifestations.

A neighbor of mine races pigeons. He takes a group of birds to some remote location along with other pigeon racers, and releases the birds to see how fast they get home. Electronic bands on the pigeon's legs register their arrival at the home loft while their owner reads the results on a distant computer. I asked how many birds get lost each race and he responded "about 40 percent." Seemed like a lot to me. In 1997, a big race was held to celebrate the centenary of the Royal Pigeon Racing Association. More than 60,000 homing pigeons were released from Nantes in northwestern France and were expected to fly to their home lofts in southern England, a distance of about 400 miles. The birds were supposed to arrive in the late afternoon, but, in the end, only about 10 percent made it—a 90 percent loss of birds! Three other races took place near the Atlantic coast that year and in 1998, all with similar losses. One explanation postulates that the Concorde, a supersonic jet that passed the area of the races while they were in progress, produced a cone-shaped shockwave that may have interfered with the birds' infrasonic hearing. There's no real proof, but it certainly sounds plausible as similar evidence indicates that noise from ship traffic interferes with communication among whales.

FLYWAYS AND THE DANGERS OF PREDICTABLE ROUTES

In North America there are four major migratory routes or flyways: the Atlantic, Pacific, Mississippi, and Central. These are general routes along the East and West coasts, the Mississippi River, and east of the Rocky Mountains, but the boundaries are fuzzy. Birds that move in family groups such as ducks, geese, swans, and cranes generally follow flyways, so the four North American routes have been useful in establishing regulations for the harvest of waterfowl. The northern Sacramento Valley of California is on the Pacific flyway, wintering grounds for millions of waterfowl and other birds. Sometimes, if I am in the right spot, as I was one year when a flock of White-fronted Geese came in for a landing, the birds came so close I could feel the wind off their wings and hear their feathers slicing through the air.

North American waterfowl flyways.

The other major flyways of the world are the Black Sea–Mediterranean, East Africa–East Asia, Central Asia, and the Australasia–East Asia. All these flyways extend into the Southern Hemisphere, but birds stop at different locations along the route. These migratory paths are considered highways along which birds travel, but birds don't restrict themselves to particular paths, so recognizing distinct belts of migration through North America and other continents is a bit unrealistic.

Songbirds wander east and west to a much greater degree than waterfowl. All we know for sure is that migrating songbirds travel between certain breeding areas in the North and certain wintering areas in the South and that certain species use a few heavily traveled corridors while other species follow more generalized routes. However, individual birds show strong fidelity to a particular route. Having traveled a certain path, a bird knows the way, the stopover points, and the hazards, and will follow the same route year after year.

These general migratory routes have evolved over millions of years, shaped by the topography of the land and water and stopover sites that provide adequate habitat, food sources, and protection from predators. Many birds wander widely but others such as the Red Knot and Purple Sandpiper restrict themselves to coastlines. Birds typically migrate south from various points across their wide northern breeding range. Ultimately the flight paths taken by individual birds and populations converge because of the constriction of the land mass and the suitability of the habitat. Eastern Kingbirds breed across a span of 2800 miles, from Newfoundland to British Columbia. On migration the width of the route they traverse narrows until the migratory path covers a width of only 900 miles from Florida to the mouth of the Rio Grande. Farther southward at the latitude of the Yucatan, it narrows even more to 400 miles.

Over the past couple of centuries, the increasing human population has had a considerable effect on birds as they follow their historical migration route. Although the United States government sets hunting limits and seasons for each species, the birds have considerably less protection once they enter Mexico and Central and South America. (There are exceptions: hunting was banned in Costa Rica in 2012, the first Latin American country to do so.) Attracted by lax hunting regulations, 7000 foreign hunters from the United States and Europe visit Argentina every year. In Bolivia I met a couple of American hunters who couldn't tell me enough about the amazing numbers of birds they killed in two days. Across the Atlantic in the Adriatic Flyway there are major abuses of bird hunting; two hunting companies actually export birds to Italy for food. On Cyprus, an island in the Mediterranean that 100 million birds visit each year during migration, 10 million migrants are killed for food by trapping with mist nets and sticky glue on tree branches. Some rest stops, which birds have adapted for their safety and resource predictability, have either disappeared or become death traps for migrants.

At least several dozen species have significantly changed their migratory routes and timing. The Barnacle Goose now breeds 800 miles south of its historical breeding site and the Sand Martin's wintering population around Africa's Lake Victoria has grown because of an increase in sand flies brought about by a reduction in fish that prey on the insects. The Eurasian Blackcap is a common bird of Europe, Asia, and northern Africa. Before the 1950s, the bird rarely wintered in Britain, preferring southern Europe and parts of Africa. By the 1990s there were thousands of birds (almost 10 percent of the European population) that bred in Europe and migrated northwest to Britain for the winter. Peter Berthold and co-researchers in Germany bred British-wintering blackcaps in captivity and when they released the

Barnacle Geese on migration.

offspring, the birds headed northwest, unlike the typical European Blackcaps. This study is probably the first to demonstrate a genetic change in migratory behavior.

Cultural evolution has influenced changes in migratory routes—some individuals of a species modify a route and the others follow. Geese, storks, and cranes, species that live for a long time, have older individuals that tend to lead the flock and may make changes in the route because of environmental alterations caused by the increasing human population. Migration patterns have been and continually are being sculpted by the environment as the survival of birds depends on their successful adaptations to changing conditions.

Years ago, on the western edge of the Great Basin desert in northeastern California, I was standing under a juniper tree with a colleague, no doubt discussing something of global significance. Above us, a throng of Tree Swallows was gathering in the branches in preparation for their southward trip. As I talked, I gesticulated with an upward-facing palm and a Tree Swallow fell in it, dead. Good grief, I thought, if one of these swallows drops dead on a nice summer day before migration is seriously underway, what are their chances for making the entire round-trip journey? The answer: slim. More than 79 percent of Tree Swallows hatched in the Northern Hemisphere die in their first year, many undoubtedly succumbing to the rigors of the round-trip passage. But those birds that endure the challenges and hazards of the trek survive to reproduce. It has to be worth it.

WEATHER SURVIVAL STRATEGIES

Enduring Heat, Cold, Wind, and Rain

October extinguished itself in a rush of howling winds and driving rain and November arrived, cold as frozen iron, with hard frosts every morning and icy drafts that bit at exposed hands and faces.

—J. K. ROWLING, *Harry Potter and the Order of the Phoenix*

Birds are at the mercy of the meteorological environment. Avian anatomy and physiology including body size, wing and leg length, beak shape, fat deposition, feather color, salt retention, respiration rates, and red blood cell count are partly the evolutionary results of climatic settings. Birds lose water from their skin and respiratory system, pant, shiver, flutter their throats or tongues, dilate or constrict capillaries, seek shade, lift a leg, spread their wings, fluff up, or otherwise respond to an environment that is at the edge of their physical comfort zone. People act similarly, but can generally seek relief offered by modern conveniences like air conditioning and furnaces.

A bird's physical environment undergoes only gradual changes unless some major disruption occurs. The composition of the air, soil, and underlying geologic substrate are stable and the photoperiod changes exactly the same way every year. The only significant variable is weather. Although birds have evolved to cope with the unpredictability of the weather to some degree, extreme events can seriously challenge their survival. The year 2012 saw an arduous breeding season for 23 out of 24 of the most common birds in Britain. Heavy rains in May and June washed away nests and made insects hard to come by, according to the Royal Society for the Protection of Birds. Only the European Blackbird saw an increase in its population, probably because it feeds on worms and other critters that normally inhabit the ground and are still accessible after a deluge. That same year, the California Audubon Society noted that after three years of a severe drought, hawks in the Central Valley showed a 90–95 percent nest failure rate because the plants that produce food and provide cover for rodents had become scarce. Penguin chicks in Argentina have suffered from multiple effects of climate change as heavy rains, strong storms, and exceptional heat take their toll. A colony of 3500 Magellanic Penguins was studied from 1983 to 2010 in Punta Tombo, Argentina. This region's climate is historically mild and dry, but 13 years of the study had an exceptional amount of rainfall as well as unusually cold weather. The soft down of penguin chicks that keeps them warm in dry weather is not waterproof and when the cold rains fell, up to 50 percent of the chicks died.

Studying birds in their natural environment is the only way to really understand how they cope with what nature throws at them. In Northern California for several winters I collected data on the territorial behavior of Townsend's Solitaires, thrushes that spend November to March in a cold juniper and sagebrush habitat at an elevation of 5200 feet. Eating juniper tree berries and an occasional larval insect, Townsend's Solitaires survive the winter by safeguarding their berry supply against intruders—other solitaires and American Robins. A solitaire would sit on a treetop whipped by the wind and call to announce its presence. Then it would drop down to the lower branches of the junipers or the ground to feed and quickly return back to the apex of a tree. After four winters of taking notes, I went to my office and crunched numbers—and this is where science pays off: the analysis of data.

The data showed that in years when the juniper berries were in short supply the birds defended the resource, so both solitaires and robins established territories, sang, called, and chased intruders. In years of abundant berries,

A Townsend's Solitaire surveying its territory.

the Townsend's Solitaires paid little attention to each other or to their robin relatives. Both species just wandered around within a loosely defined area eating berries *ad libitum*. Energy expended in defending a territory when there is sufficient food only puts a bird at risk for accidents or predation. But when food is scarce—especially in the winter when any morsel of energy can mean the difference between surviving the night or not—it is worth protecting. These kinds of strategies and struggles go on all the time but weather overlays every challenge to survival.

REGULATING BODY TEMPERATURE

Humans are comfortable at certain temperatures, but other temperatures make us shiver or sweat. That range of ambient temperatures that don't make us sweat or shiver is our thermoneutral zone (TNZ), in which our basal metabolic rate (BMR) is stable and biochemical processes are optimized. Above and below the TNZ, birds and mammals have to expend additional energy to keep cool or warm. We are able to acclimate physiologically (to some degree) to different ranges of temperatures, depending on where we live. If you live in North Dakota a temperature slightly above freezing might seem mild, but Californians experiencing that same temperature will be pulling out their down parkas and earmuffs. Birds too can acclimate somewhat, so their TNZ changes with the season, latitude, and local conditions. In general the TNZ spans a greater range of temperatures in colder environments than in warmer ones and in larger birds compared to smaller ones. For example, the TNZ of the tiny Green-backed Firecrown hummingbird of Chile and Argentina is 55–82°F versus 50–122°F for the Ostrich of mid- and south Africa, which weighs as much as two average humans. The TNZ of the 6 ounce desert-dwelling Gambel's Quail of the southwestern United States is 86–111°F, and the Adélie Penguin of the Antarctic coast is 14–68°F.

We say that birds and mammals are warm-blooded, but that's a misleading term. The desert iguana, sitting on a rock in the sunshine of a desert can absorb enough solar energy to bring its body temperature up to 115°F. Even though that would make the lizard warm-blooded, lizards are considered cold-blooded animals. The proper term for warm-bloodedness is "homeothermic," meaning that the body temperature remains relatively constant regardless of the ambient temperature; this enables birds and mammals to exploit habitats with extremes of temperatures. Homeothermy requires a cornucopia of mechanisms to maintain the stability of the body's

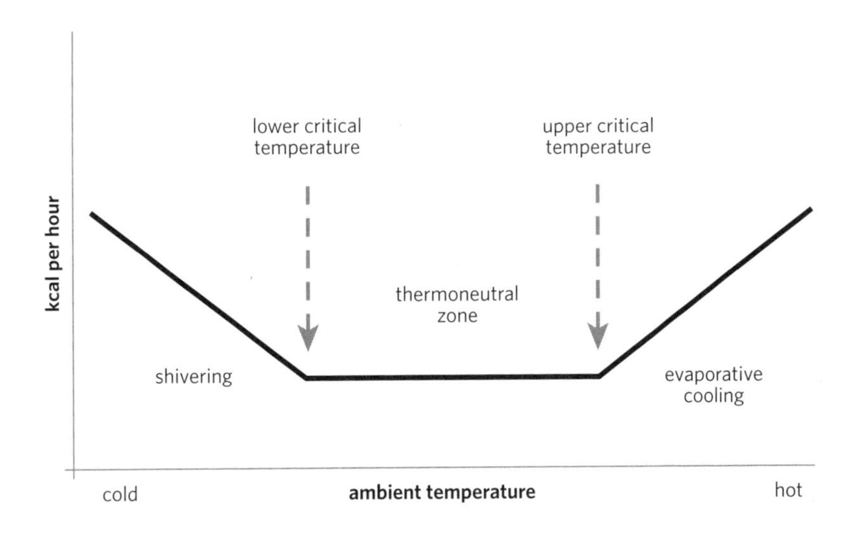

Every bird species has its own thermoneutral zone, which provides a major explanation for the geographic distribution of bird species.

internal environment. In the winter homeotherms need to produce more heat because more heat is lost to the environment in cold temperatures. To produce this heat, the BMR has to increase. Studies of small birds in winter environments showed that the BMR of Black-capped Chickadees increased 6 percent versus a 17 percent rise for Mountain Chickadees and 22 percent for the Juniper Titmouse. The body temperature of cold-blooded organisms, on the other hand, changes with the ambient temperature; they are "poikilothermic," meaning varying temperature. Poikilotherms either cannot inhabit a cold environment or do so in fewer numbers because their BMR would be so low as to preclude activity. Very few amphibians and reptiles live above the Arctic Circle and none exist in Antarctica.

But even homeotherms have their limits. Although variables such as food, predators, nesting sites, and competitors constrain the population of birds in particular geographical areas, it is temperature (and other meteorological factors to a lesser degree) that fundamentally controls the range of the entire species.

Bergmann's Rule and Allen's Rule

Warm-blooded animals tend to have larger body sizes in colder regions; this concept is called Bergmann's Rule. As body sizes increase, an animal's volume increases faster than its surface area. Because the bodies of mammals and birds generate heat and the surface area loses heat, larger animals generate proportionately more heat and lose proportionately less heat than smaller animals. Sometimes overheating is a problem; this is why elephants have big ears—to radiate heat. It also explains why hummingbirds are not typically found in cold environments—their small bodies lose heat too quickly for their metabolism to keep up. Meiri and Dayan of the University of Tel Aviv examined 94 bird species worldwide and determined that 72 percent of them adhere to Bergmann's Rule, although a higher percentage of sedentary species reflect the rule than migratory ones. Body sizes of non-migratory Downy Woodpeckers in the northeastern United States, for example, are larger by almost 10 percent than those found in southern Florida. Similar patterns hold for the sedentary Blue Jay and White-breasted Nuthatch. It makes sense that any bird species spending all its time in one environment should be more closely attuned to that environment than a species that spends only part of the year there. Not surprisingly, 75 percent of bats also adhere to Bergmann's Rule.

Allen's Rule says that the appendages (arms, legs, ears, snouts) of homeotherms are shorter at higher latitudes and colder temperatures because longer appendages mean higher heat loss. Cartar and Morrison of the University of Lethridge and the Canadian Wildlife Service examined 17 species of breeding shorebirds in the Canadian Arctic and determined that the lengths of tarsi decreased as the harshness of the environment—as judged by wind speed, solar radiation, and temperature—increased. They argue that birds with longer legs are more exposed to the wind's cooling effect and thus have to spend more energy keeping warm. Similarly, Nudds and Oswald of the University of Leeds in the United Kingdom found that the unfeathered portions of the legs of 43 species of terns and gulls are shorter in colder environments.

New evidence for Allen's Rule came about after the discovery of the thermoregulatory capability of the toucan bill. (I'll get to that shortly.) University of Melbourne scientists measured the beaks of 214 species of birds from across the world and found a relationship between bill size and latitude—the higher the latitude, the smaller the bill. The role of bird bills in heat regulation had received scant attention with research instead focusing on bill configurations as related to feeding habits, so this study was significant. The researchers concluded that there is a thermoregulatory cost to a larger bill so selective pressures produced smaller beaks in colder climates. Thermal imaging demonstrated that bills with a larger surface area dissipated 33 percent more heat than smaller bills. Song Sparrows living in drier habitats have bills with larger surface areas than Song Sparrows in moister habitats, which tend to be cooler. The only exception the researchers found was among the grass finches of Australia. Apparently, the finch bill's seed-handling capability is under stronger selective pressure than its thermoregulatory costs.

Adaptations for survivability are not always neatly organized and it's worth noting that Allen's and Bergmann's Rules are merely rules of thumb for birds. Many exceptions exist such as the Yellow Warbler, whose breeding range extends from the Caribbean to the Arctic, and whose characteristics do not support either rule.

DEALING WITH HEAT

Heat stress (hyperthermia)—elevated body temperature—appears, overall, to be more severe than cold stress (hypothermia) in birds. Overheating is a common problem faced by poultry producers, as too high an ambient temperature causes a loss of body and egg weight and poor quality meat. But overheating is also a concern for wild birds, more so now with increasing global temperatures. In January 2009, thousands of birds, mostly parrots, died in Carnarvon, Western Australia, when the temperature reached 113°F; the following year 208 Carnaby's Black Cockatoos, an endangered species, were found dead when the temperature hit 118°F. A small bird can lose 5 percent of its weight every hour at that temperature, leading to dehydration and eventually death in a matter of hours.

Birds have many morphological and physiological adaptations that allow them to respond to a range of temperatures and survive with little or no stress. As the ambient temperature increases, physiological mechanisms approach their limits and behavioral mechanisms come into play. Here are some of those adaptations and mechanisms for surviving the heat:

BILLS THAT RADIATE HEAT The toucan's bill has always been fascinating but it was made even more so when a new role for the Toco Toucan's bill was revealed in 2009 via infrared thermography. With a large and colorful bill used for picking fruit and capturing small animals, this tropical South American bird also employs its beak to radiate body heat. The surface area of the bill covering is about half of the body's total surface area, uninsulated, and well supplied with superficial blood vessels. While flying, the toucan might increase its metabolic heat production by a factor of 10–12, so in a warm tropical environment the bill is an important source of heat loss. Another indication that the bill radiates heat is the toucan's common practice of tucking its bill under a wing and folding its tail forward over the bill while at rest to reduce heat loss. Relative to its body size, the Toco Toucan has the largest radiator in the animal world, even compared to the ears of elephants.

The Toco Toucan's bill was discovered to be a radiator.

EVAPORATION VIA THE SKIN When we sweat, moving air wicks the water off our skin to provide a bit of cooling. Birds do not have sweat glands but their skin contains water and cutaneous water loss assists in cooling; how much cooling depends on the thickness of the skin, the amount of fat, and the blood supply. In birds, the fat (lipid) layers of the skin minimize water loss in a thermoneutral environment, but also function to facilitate heat loss during episodes of hyperthermia. Verdins, like many small birds, stay within their TNZ by losing water, along with accompanying heat, from their skin. This cutaneous evapotranspiration accounts for 50–65 percent of the bird's total water loss. However, as the temperature rises, respiratory water loss—from lungs and air sacs—becomes increasingly important. Diamond Doves of the arid deserts of central Australia have little access to water. They feed in the heat of the day without shade cover but to survive they conserve water by allowing their body temperature to rise, like camels, but remain within the TNZ. The more heat that can be stored, the less water is needed to dissipate it. As the air cools later in the day, the bird will need to use a lot less water to dispose of the excess body heat.

AVOIDING OVERHEATING As birdwatchers know, early in the morning and late in the afternoon are far better times to watch birds than midday. Near the equator in places like Bolivia or Uganda, little or no bird activity occurs in the hours around noon as the heat and humidity make it nearly impossible for a bird to engage in any serious movement. Conversely, in cold environments, birds are more active because they need more food to survive and the energy they expend in foraging helps to generate body heat.

PANTING AND GULAR FLUTTERING Panting, the rapid increase in respiratory rate and air volume, is an additional mechanism for heat loss in birds subjected to heat stress. Songbirds lose excess heat by cutaneous water loss as long as the birds are sufficiently hydrated. When dehydration occurs, loss of water from the respiratory system comes into play, and if that isn't sufficient, panting begins. Panting movements may be as much as 16–27 times the normal respiratory rate. If the heat load continues to increase, some non-passerines (non-songbirds) may also employ gular flutter. In these birds the gular or neck region of the skin from the bottom of the jaw to the throat is featherless or nearly so. The floor of the mouth and throat vibrate by the use of hyoid bones and muscles, helping to dissipate heat from the bare skin. Gular flutter is common in seabirds such as pelicans, cormorants, anhingas, boobies, frigatebirds, and ground-dwelling birds like turkeys, pheasants, and roadrunners. These large birds generate considerable body heat for which overheating could be a problem without gular fluttering. The flutter rates of various bird species vary from 235 to 735 flutters per minute. Panting and gular fluttering may be synchronized as they are in owls and domestic pigeons or uncoordinated as in pelicans and cormorants. The difference appears to be that the gular region of birds that show synchronization is comparatively small and may simply be driven by the movement involved in panting while unsynchronized panters have large gular regions that are controlled independently of panting. The energetic cost of panting is high, as is the amount of water lost, so birds may use the more efficient mechanism of gular flutter instead to conserve both energy and water. Monk parakeets employ lingual flutter, vibrating their tongue in coordination with their respiratory rate. The cloaca, the common urogenital opening, can also be used to dissipate excess heat; this is probably an emergency measure as it only occurs at particularly high ambient temperatures. Cloacal evaporation in Inca Doves has been measured at 21 percent of total evaporation at a temperature of 107.6°F.

SEEKING SHADE When we get too warm we can seek shade, fan ourselves, find a breeze, drink water, or jump in a pool—birds can employ similar behaviors. Desert-dwelling birds probably face the greatest challenges. Rock Doves in the Negev Desert move among rock formations to seek shade on hot days; when they sit, they spread their wings and erect their dorsal feathers to dispel heat from their backs. Birds in arid areas like the Sonoran Desert minimize heating and water loss by installing themselves in tree crevices during the hottest part of the day, forgoing foraging activity to decrease their exposure to high temperatures. The Crowned Plover, which lives in eastern and southern Africa in short grass habitats, raises itself on its legs above its nest and spreads its wings to expose more body surface area to the air. Although the bird shades and cools its eggs while doing this, the primary

Gular flutter in a cormorant.

point of this behavior is to disperse heat from the bird's body. The Hoopoe and Crested Larks of the Arabian Desert (where more than a year can pass without rainfall and the average summer humidity might reach only 15 percent) confront air temperatures of more than 113°F and ground surface temperatures as high as 140°F. To avoid the heat of the day, the larks forage in the early morning and late afternoon and spend the rest of the day resting in the shade of shrubs. During the hottest times of the day, the birds often hole up in the burrows of Egyptian spiny-tailed lizards, scrape away the top layer of sand, and prostrate themselves with their neck and chest against the ground to conduct heat away from their body—this reduces their potential water loss by 81 percent. The birds are capable of lowering their basal metabolic rates by about 40 percent to decrease the loss of water from their skin and respiratory systems. Early in the nesting season while it is still relatively cool, most lark nests are built on the open ground, but as the nesting season progresses, the majority of nests are built in or under shrubs.

RAISING OR FLUFFING OF FEATHERS Birds use the technique known as ptiloerection to both increase and decrease heat loss, the difference being in the position of the feathers. To increase heat loss, the body feathers are

A Crested Lark in the cool early morning hours.

lifted high enough so that the skin is exposed and cutaneous water loss increased for cooling as the blood flow to the skin is increased. Australia's Spinifex Pigeon lives in an arid environment where half of the year the temperature in the shade exceeds 100°F. The bird survives by having a low metabolic rate and manages its cutaneous heat loss by elevating its skin temperature and raising its feathers. Great Knots wintering on the north coast of Australia prepare for their long migratory flight to the Arctic by adding layers of fat, but because they are doing this preparation in a warm environment they face overheating. To avoid heat stress, the birds raise the blackish feathers of their back to reduce absorption of solar radiation and increase both convective and cutaneous cooling. An alternative to raising feathers is flattening them, as the Curve-billed Thrasher of the southwest United States and Mexico does. Compressing the feathers against the body reduces their insulating value by eliminating most air spaces. In addition to gular fluttering, the Great Frigatebird engages in three heat-dissipating postures. When heat stress is minimal, the birds erect their feathers on various parts of the body; as the temperature increases, they droop their wings; and during the hottest temperatures, they spread their wings for maximum cooling. European Bee-eaters at the southernmost tip of Israel have been observed diving into salt ponds and the Red Sea. After emerging from the cooling water, they perch in a tree and spread their wings to take advantage of any breeze and the low humidity. Occasionally birds get waterlogged and have to lie on the beach with outspread wings to dry before they can fly. Besides that inconvenience, bee-eaters have also been found in the stomach of sharks.

UROHYDROSIS Wetting the legs by excreting on them is sort of like spraying your legs with water and letting evaporation cool you off. The Turkey Vulture spreads its wings, extends its neck and head, and, engaging in a behavior that adds to the bird's rather unpleasant reputation—it poops on its legs. An interesting problem was discovered several years ago when a bird bander found that several of the Turkey Vultures that he had banded developed lesions on their lower legs because of the accumulation of waste matter on the leg bands. (The U.S. and Canadian banding offices no longer allow leg bands on vultures.) Urohydrosis is rather rare in the bird world because it requires easy access to drinking water, but New World vultures, herons, condors, storks, gannets, and boobies also engage in this unsavory activity.

PLUMAGE COLOR White-feathered birds reflect more solar radiation than darker birds so we expect white birds to be more common in hot environments. Herring Gulls nesting in the open are subject to intense solar radiation, so when they not able to leave the nest when incubating or brooding, they orient their bodies so that their whitest plumage, with the highest reflectance, faces the sun. So why aren't all birds in hot environments white or at least a light color? The size of the bird, the flight functions of the feathers, the microstructure of the feathers and skin, the range of ambient temperatures, the bird's migratory habit, the plumage's social or sexual signals, and visibility to predators are all factors that could affect the color of a bird's plumage. To complicate things further, Gloger's Rule says that birds and mammals living in humid environments tend to have darker plumage or fur than animals of the same species living in drier habitats. For example, Song Sparrow populations of the Pacific coast show increasingly darker plumages from southern California deserts north to the moister and cooler Pacific Northwest forests. The same pattern of south to north darkening is seen in House Sparrows in the Rocky Mountains and the Midwest, and in more than 90 percent of the bird species of North America. One explanation is that melanin, the black-brown pigment, makes feathers stronger and more resistant to degradation by feather bacteria, which grow well in humid environments. Also, perhaps, the paler colors of birds in drier and lighter-colored environments help in concealment.

SURVIVING THE COLD

People in cold environments, unlike birds, can take advantage of a variety of clothing to keep warm—mukluks, parkas, sweaters, and thermal underwear. It makes me cold just watching geese sitting on frozen ponds, shorebirds pecking away on mud flats in a biting wind, and gulls flying through heavy rain over winter ocean swells, but it appears that they are more physically, physiologically, and behaviorally suited to avoid hypothermia than hyperthermia, and not just because some of them migrate to the tropics for the winter.

George Murray Levick was a surgeon and zoologist on Robert Scott's disastrous expedition to the Antarctic in 1910–1913. Scott did not survive but Levick and his notebooks did, in which he described the sexual behavior of Adélie Penguins. He was so taken aback by their "astonishing depravity"—homosexuality, coercive sex, and sex with dead females—that he wrote his observations in Greek so that only "learned gentlemen" could understand. Some of that depravity results in baby penguins, such as those in *March of the Penguins*. The film's depiction of fuzzy young penguins being whipped by the blistering winds of the Antarctic makes one wonder "how do they make it?" It takes about 15 days after hatching for young Gentoo and Chinstrap Penguins to develop a sufficiently insulating layer of down, so they need to be brooded by their parents for that time; when the offspring are 25 days old, their downy coat is quite adequate for the task. The feathers are stiff and short, and not arranged in tracts like other birds—instead the feathers cover the entire body, overlapping like roof shingles. The wind, rather than ruffling the feathers, compresses them, increasing the young birds' insulation. In fact, when the winds bluster over the colony, the young are comfortable because the insulating efficiency of their coat is not only ample, but 136–178 percent better than in calm winds! The contour feathers of the adults are similarly compressed when the birds are swimming, increasing both insulation and aerodynamics. In addition, the deeper feather layers are composed of ever-smaller feathers that produce insulating pockets of air.

As winter approaches, many birds increase their body fat for energy storage and insulation and thicken their plumage with extra down feathers. The Common Eider, a northern sea duck, increases the insulation value of its plumage by 25 percent for the winter by adding down. When molting in the breeding season the eider uses its down feathers for nest lining. Vikings learned to make down comforters out of eiderdown for their long voyages. Today the only eiderdown harvested from wild birds is in Iceland, where it is collected from nests after the breeding season and a small truckload exported annually, which is why eiderdown comforters are so expensive. The Rock Ptarmigan, a grouse of the Subarctic and Arctic Eurasia and North America, puts on extra fat, comprising up to 32 percent of its winter body mass, almost doubling its weight; that fat plus down results in an insulation-efficient body. American Robins might increase their feathers by 50 percent as winter approaches; American Goldfinches nearly double their body feather count.

I was once asked about pregnant birds. After I thought a bit, I realized the inquirer was looking at a Hermit Thrush that had ruffled its feathers, giving it an apparently larger body. Ptiloerection, used for cooling, is also employed in reducing heat loss. How many holiday greeting cards have you seen with a fluffed up bird pictured against a snowy background? A moderate amount of fluffing traps sufficient air to form layers of feathers and increases insulation value by 30–50 percent. However, on a rainy day, this might not be the best strategy. Laboratory studies of American Kestrels demonstrated that fluffing during a rainstorm allows water to penetrate to the skin and lower their body temperature, so the birds slick down their plumage instead.

Instead of adding extra feathers for cold weather, Eurasian Bullfinches increase the amount of fat ingested, not for insulation but for food reserves. Birds need to eat enough food during the day to put on enough fat to allow them to survive the night. The colder the temperature, the more foraging required to accumulate fat, but there is a tradeoff between expending the energy necessary to accumulate fat stores and being more exposed to the weather and predators while feeding. Heavier birds are also more at risk from predators as they move a bit more slowly. Every day is different because of variation in the weather, competitors, predators, and the food supply.

Hermit Thrush fluffed up for warmth.

One strategy for minimizing energy expenditure and maximizing food resources is to establish a territory. Both male and female European Robins hold winter territories. The male holds the same territory all year round and the female establishes one to protect a winter foraging area. Singing to advertise their protected area is important but it takes energy. Robins with a higher body mass sing more when they greet the dawn than thinner birds. The birds were less apt to sing at sunrise if a cold night required them to use their fat stores. Similarly, the amount of time nightingales spend singing during the night is directly related to their latitude (higher latitude, lower temperature, less singing).

Countercurrent Heat Exchange

Losing body heat through respiration is handy in a hot environment, but not so helpful when it is cold, so physiology has had to adapt. When we, and birds, inhale cold air, it is heated as it passes through our nasal passages, throat, trachea, and bronchi. When the air reaches the lungs it is almost body temperature; then we exhale warm air. In birds, a countercurrent exchange of air occurs in the nasopharyngeal passages so that some of the heat from the

Sanderlings standing on one leg.

expiring air is recaptured. Depending on the ambient temperature and the species of bird, the amount of heat recaptured can be as high as 80 percent, as it is in penguins.

The surface area of a beak may be a compromise between heat loss and foraging requirements. Likewise, leg and toe lengths are compromises between locomotory needs and heat loss. The old joke I lay on beginning birdwatchers is "Why do some birds stand on one leg?" The answer: "Because if they lifted it up they would fall down." After the groans subside I explain that to reduce heat loss in cold temperatures, some birds like the Sanderling tuck one leg up into the feathers of their abdomen while resting, or both feet while sitting down or floating on the water. But it is not always possible to tuck one or both legs away. Instead, ducks, gulls, tinamous, and many other birds have a countercurrent heat exchange system between the arteries and veins of their legs. Blood in the arteries from the heart is warmer than venous blood returning from the extremities. The arteries and veins are intertwined in a structure in the lower leg called the *rete tibiotarsale* (net of the tibiotarsus) which may consist of as few as three arteries and five to seven veins, as in owls, or up to 60 arteries and 40 veins as in flamingos. Birds without this net have two veins, one of which runs closely on either side of an artery. The warmer arteries pass on some of their heat to the colder veins, returning some warmth to the heart but still providing enough heat to the legs and feet to keep them from being frostbitten. In some species of gulls and guillemots, the arteries and veins are closer to each other in the birds of more northern areas than in those of more southerly locations.

Countercurrent heat exchange operates very efficiently. A duck standing on ice will lose body heat, but only 5 percent of that loss will come from its feet. Of course, blood has to supply oxygen and nutrients to the feet and legs, but only a small amount of blood is needed because the muscles that operate the feet and legs are mainly concentrated in the upper leg and utilize long tendons for mobility. At lower temperatures that threaten the feet with frostbite, blood flow is increased to the lower limbs by opening special valves in the arteries. By the same mechanism, birds can lose body heat in a hot environment by shunting more blood to their legs and feet.

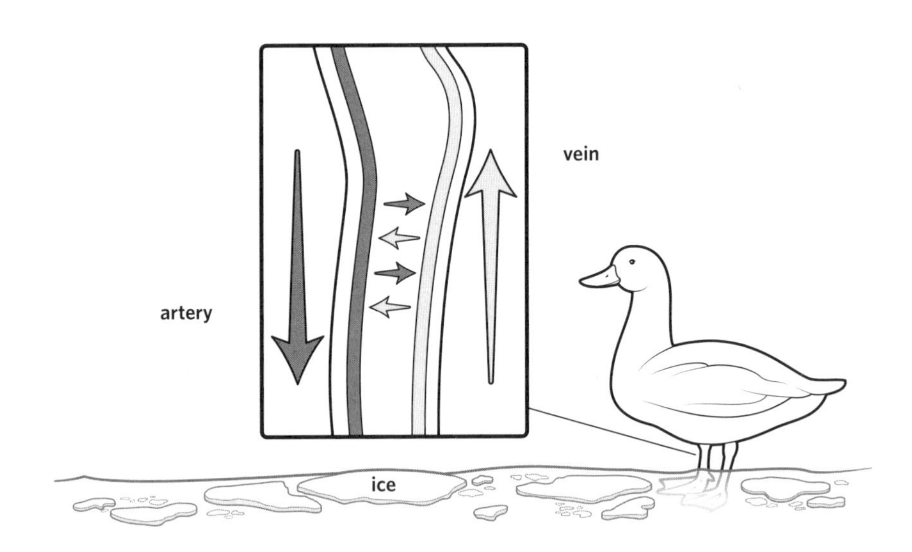

Heat exchange in lower leg. In countercurrent heat exchange, blood flowing in the arteries toward the foot is warm; the blood gives up some of its heat to the veins going toward the heart.

Huddle Pyramids and Other Behavioral Adaptations for Cold

Physiological and physical adaptations diminish the effects of cold, but behavioral adaptations can be just as effective. Penguins of the Antarctic are exposed to temperatures as low as -94°F but yet their feet do not freeze. Again, countercurrent exchange, minimal muscle in the feet, and feathered legs all help, but the birds might also squat down and let their abdominal feathers droop over their feet. Penguins can also lean back on their heels and tails and lift the front of their feet up off the ice and against their body. Loons and grebes, spending most of the day on the water, extend a leg, shake the excess water off, and tuck one foot or both feet under a wing. Hummingbirds crouch so their downy belly feathers cover their feet. Some sparrows and other songbirds drop to the ground and hunker down. Shorebirds, gulls, ducks, herons, and other birds may rest with their bills tucked under a wing or scapular feathers, mimicking what we do when we wrap a scarf around our faces. Infrared studies of heat loss indicate that one of the areas of greatest heat loss in birds is around the eye, one reason why a cold bird, especially a small one, tucks its head under a wing.

Birds huddle together to conserve body heat. Flocks of Pygmy Nuthatches, sometimes more than 100 birds, spend the night close together in a tree cavity. Dozens of Tree Swallows survive snowy nights in Canada by perching close together on a wire while tucking their beaks into their feathers. Typically solo foragers, Snow Goose goslings spend more time huddling together when the ambient temperature drops. European Long-tailed Tits roost tightly together in a line at night, competing for the middle positions as the birds on the ends get colder and lose more weight overnight. Bald Eagles prefer the middle of a roost and the colder the ambient temperature, the more birds seek the middle, although the maintenance of dominance hierarchies has a role as well. Inca Doves will stand on each other's backs, up to twelve birds three layers high, to form a huddle pyramid for warmth. Penguins too group together tightly in the chilly Antarctic environment. With temperatures down to -60°F and winds up to 100 mph, the temperature in the middle of the penguin pack might reach 70°F. The birds continually shuffle around so that penguins on the edges move toward the middle and vice versa and the entire group survives by sharing heat loss.

Being wet adds to the problem of retaining body heat. Anhingas (also called snakebirds, darters, and water turkeys) are found in many shallow warm water habitats of the world. They have wettable feathers, allowing the birds to make quick dives. Although they have fairly low metabolic rates, their aquatic lifestyle causes a loss of body heat, so when they leave the water they perch in a wings-spread posture with their backs to the sun, both to dry their plumage and absorb solar radiation. Bank Cormorants, endemic to the cold ocean waters of Namibia and western South Africa, allow their body temperatures to drop to nine degrees below normal while foraging continuously for nearly an hour. Like anhingas, they recover their body heat by basking in the sun.

A population of Great Cormorants winters in Greenland near the Arctic Circle, with water temperatures near freezing and sub zero air temperatures. They live there even during the polar night (24 hours of darkness) when they

Emperor Penguin chicks huddling.

are unable to sunbathe after foraging. How do they survive? Great Cormorants are the most efficient foragers of all water birds ever studied, capturing 0.6 ounces per minute, or about 2.2 pounds of fish per hour, so they only need to spend 2 percent of their daily activities fishing and 3 percent of their time flying. Eating plenty of fish and resting most of the day keeps their body temperature within the TNZ. Even when the sun is shining, they do not sun themselves with spread wings. It is so cold, humid, and windy that if the birds spread their wings to dry, they would lose body heat. On the other end of the world, the related Antarctic Blue-eyed Shags behave similarly.

Lots of birds sunbathe, though. African Penguins (formerly "Jackass Penguins" because of their braying call) orient their black backs and flippers to the sun soon after sunrise for an hour or two before becoming active. Herons, egrets, pelicans, vultures, hawks, and hawks engage in sunbathing, but we can only assume that this behavior is to absorb solar radiation when the ambient temperature is low and/or the birds are wet. Other reasons for wing spreading include losing heat, drying out, or allowing the sun to kill skin parasites and bacteria. An experiment in which both louse and mite populations in a colony of Violet-green Swallows were reduced with the application of pesticides resulted in the birds sunning themselves considerably less than birds in a colony of swallows that was not sprayed. Birds get some of their vitamin D from sun exposure, but since the sun can't reach their skin, they use another mechanism. The precursor chemical for producing vitamin D is found in the uropygial or preen gland. During preening the chemical is spread on the bird's feathers and when exposed to ultraviolet light, changes to vitamin D. The next time the bird preens, it ingests the vitamin D. Water birds like the Shoebill Stork preen more often to stay waterproof.

The unusual Shoebill Stork of Africa, ugly as it may be, still needs to preen.

Shivering and Non-Shivering Thermogenesis

When you spend too much time in the cold, you began to shiver—the body's physiological response via small contractions of mostly skeletal muscles to produce more heat. Birds do the same thing; this is shivering thermogenesis. Most of this energy comes from fat stores and much of the shivering occurs in the pectoralis muscles. As winter approaches, birds acclimate to the increasingly colder weather and don't shiver as often, although in severe weather, survival might require shivering for days. Overwintering birds obviously have the greatest need for the extra heat produced by shivering, but shivering also occurs in some migrants traveling north from the tropics. One study of a vireo indicated that on spring migration the bird had the capacity to generate 17 percent more heat from shivering than it did during the summer. On their northward migration, birds move rather quickly to their breeding grounds but may run into cold weather along the way, so this extra heat-generating capacity might come in handy.

Non-shivering thermogenesis is the generation of body heat without muscle contraction. The digestion and absorption of food requires energy and produces body heat. The Tawny Owl of the northern forests of Eurasia is nocturnal, feeding primarily on small birds and mammals. These feeding bouts produce enough body heat so that the bird generally avoids shivering even on the coldest nights. A bird can maximize its heat generation by "choosing" the best schedule for food intake. Instead of using energy to generate heat, some birds, such as the Gray Jay of the Arctic, enter a state of hypothermia by letting their body temperature, normally 107°F, drop to about 99°F; this is torpor.

Torpor and Hibernation

It took awhile to establish the basic facts about hibernation that we now know. Twenty-four centuries ago, Aristotle watched summer swallows as they flew over a marsh and dipped into the water's surface. Noticing their absence from Greece in the winter, he deduced that the birds must have dived into the water and buried themselves in the mud below to spend the colder months. Sure enough, next spring the swallows were flying above the water. This speculation spread widely and for many years fishermen reported that birds emerged from their nets along with fish and that chunks of mud dredged from the bottom of a swamp contained hibernating birds that flew off after they were released from the muck and warmed up. This bizarre myth steadfastly persisted into the early 19th century. Others who witnessed the disappearance of birds speculated that they hibernated underground.

Nothing as spectacular as mud hibernation occurs in the bird world, but at least 29 species of birds demonstrate a lowering of their metabolism and body temperature for periods of time. If a hypothermic bird exhibits a significant depression of metabolic function and little or no response to external stimulation, it has entered torpor. German researcher Elke Schleucher, in her review of avian torpor, tells us that the major reasons for entering that state are a lack of food and cold weather. Torpor has arisen over evolutionary time as a sort of emergency measure for many birds facing low temperatures and/or temporary food shortages and for a few birds as a way of saving energy to make it through the night. Some think that birds might use torpor during migration to reduce the amount of foraging necessary en route.

Virtually all birds that enter torpor are insectivorous, frugivorous, or nectarivorous because their food sources are so unpredictable. In a laboratory study of Red-backed Mousebirds, researchers found that if the birds were deprived of food until they lost 35 percent of their body weight, their metabolism would drop to one-third of normal and the birds would enter torpor. Torpor in free-living mousebirds is rare, however, as they cluster together to share body heat on cold nights. The normal daytime body temperature of hummingbirds is about 100–104°F, but during nighttime torpor, their body temperatures can decrease to the ambient temperature if it is not too cold. A couple of hours before dawn, with no apparent stimulus except its innate circadian rhythm, a hummingbird will begin to arouse from torpor. Respiratory and heart rates increase, the bird vibrates its wing muscles, shivers, and after 20–60 minutes the bird is ready to begin feeding. The mainly insectivorous female Puerto Rican Tody may exhibit torpor even with abundant food and tropical conditions, perhaps because of the stresses of breeding, and can lower her body temperature by 25°F although she remains awake and alert.

At least one bird goes into such a deep state of torpor it resembles hibernation. The Poorwill was called *holchko*, the sleeping one, by the Hopi Indians.

Lewis and Clark may have found one of these birds in a state of torpor in 1804 although a confirmed sighting was not until much later, in 1879 California. The Poorwill is a robin-sized insectivorous bird of the American southwest; all its relatives—nightjars, potoos, and nighthawks—enter torpor, but the Poorwill is the only one to do so for extended periods. Torpor is used extensively by Poorwills when the ambient temperature is below 50°F. Typically roosting under a cactus or next to a rock while facing a southerly direction, they might awake if they are warmed by sunlight and spend the day foraging, only to reenter torpor at night. Some Poorwills spend 10–20 days in a state of torpor, with no stirring at all. This behavior is very much like that of hibernating small mammals that wake occasionally to snack on their stored food cache. There has been speculation for many years that some swifts and swallows hibernate but most likely they are entering a moderate level of torpor.

HOW WEATHER AFFECTS BIRDS: MIGRATION, FOOD SUPPLY, AND DISEASE

The Old Farmer's Almanac says "If crows fly in pairs, expect fine weather; a crow flying alone is a sign of foul weather." No evidence supports this particular saying, but some bird behavior can be used to predict bad weather. Have you ever noticed birds, often swallows, perching on power lines as a storm approaches? That's because birds perch more as a low-pressure center (cold front) approaches. Low barometric pressure is a reflection of the reduced

Common Poorwill sitting on a gravel road.

density of air molecules that makes it energetically more expensive to fly. In one study, captive White-throated Sparrows were experimentally subjected to different pressure regimes. Under high pressure the birds awakened in the morning and began to preen themselves in preparation for the day's activities. When the pressure was lowered, the birds awoke and immediately began to feed, expecting bad weather. Apparently a sensory organ in the middle ear can detect changes in pressure because of shifting weather conditions as well as altitude, but little is known about that mechanism.

Every day the weather is different and birds have to face extremes of temperature, wind, rain, snow, and solar radiation and try to stay in their zone of thermoneutrality. When they cannot, nature has provided birds with an array of adaptations to survive until they again reach TNZ.

Weather-Related Mortality of Migrating Birds

The impetus to migrate is not the weather, although weather unquestionably affects the migratory behavior of birds. Heavy rain, hail, snow, and wind can take a heavy toll on migrating birds. Storms can delay departure and winds can slow or speed up their passage or blow birds off course. A snowstorm interrupted a spectacular migration of Lapland Longspurs in northern Iowa and southern Minnesota on March 13–14, 1904. It was not a particularly cold night and the wind was calm, but heavy wet snow was falling. Birds fell from the sky and landed everywhere—in towns, on the roads, on roofs, and

Bank Swallows gathering before a storm.

especially around streetlights. Floating on the surface of two small lakes in the area were an estimated 750,000 birds. The total estimate for this disaster was 1.5 million dead Lapland Longspurs.

If it is raining, feathers can become saturated and migrants might have to interrupt their journey. Birds may have the option to stop if they are traversing land but over water many birds would die. Land birds passing over large bodies of water are often lost without a trace, drowned or eaten by scavengers on the water or after being washed ashore. An enormous loss of migrating birds, estimated at 40,000 migrants of 45 species, occurred during a tornado on April 8, 1993, off of Grand Isle, Louisiana, a barrier island at the northern edge of the Gulf of Mexico that is reputed to have the highest density of birds in the United States during a spring migratory landfall. Indigo Buntings appeared to have suffered the greatest losses, but Cerulean and Swainson's Warblers and the Seaside Sparrow also showed significant mortality. Water birds such as ducks, geese, and grebes don't face an overwater problem as they can rest on the water but may have the opposite problem while crossing over large expanses of land. Eared Grebes winter in the southwestern United States to southern Mexico, but have to cross large expanses of desert to get to their nesting sites in the western United States. Even if the weather is reasonable, strong headwinds can prevent the migrants from reaching their destination, causing the birds to expire from exhaustion or starvation. Small songbirds are most vulnerable to the vagaries of the weather, but one incident of strong storms passing over the Mediterranean killed at least 1300 birds of prey. Neither are seabirds immune from ocean storms. In the first two months of 2014, more than 20,000 seabirds, mainly puffins, auks, and murres, died in the seas off the Atlantic Coast of France, primarily because of starvation and exertion as the birds tried to avoid storms and seek food. According to the French Society for the Protection of Birds, this was the largest die-off of birds in France since 1900.

Hurricanes, low-pressure storms with winds exceeding 74 mph, do not usually seriously affect birds even though they often occur in the Atlantic during the peak of migration season, mid-August to late October. Generally, migrating birds do a pretty good job of avoiding or evading storms. But birds die, of course, some from hypothermia because of wet plumage, especially young chicks still in the nest. High winds alone can kill birds if they are hit by flying debris, thrown into objects, or blown so far off course that they become disoriented and exhausted. Hurricane Sandy in 2012 produced some unusual sightings such as Northern Gannets, normally found in the North Atlantic, in New York Harbor, and a Pomarine Jaeger off Cape May, New Jersey, far from its usual migratory path in the Atlantic. A 2005 study in Quebec found that the local Chimney Swift population fell by 50 percent after Hurricane Wilma blew the birds astray, some as far as Western Europe. (These swifts are typically only found in the eastern part of the United States while breeding, and the western part of South America while wintering.) In 1989 Hurricane Hugo hit the shores of South Carolina with winds of 87 mph. An analysis of the bird populations afterward, albeit rather crude and incomplete, indicated only minor mortality. Seabirds moved northward before the storm and land birds stuck close to the ground. A study of the birds of a Puerto Rican forest immediately after Hurricane Hugo revealed that nectarivores and frugivores declined precipitously while omnivores and insectivores increased and granivores only slightly declined. Less than a year later, the numbers returned to their pre-hurricane abundance, indicating that the mortality rate was low—the birds just temporarily moved elsewhere.

The major effects of hurricanes (as well as tornadoes, fire, flood, and earthquakes) are more significant if the habitat is significantly altered. Normal food sources might be gone, nest sites or protective tree cavities damaged or destroyed, and there might be increased exposure to predators. For any creature acclimating to a new environment, the first survival tactic is to find alternative food sources. Before Hurricane Hugo, the Antillean Euphonia, a colorful tanager-like bird, specialized on mistletoe berries; after the hurricane, it ate berries from at least eight other plants. Typically, for a year or two after a hurricane, the reproductive success of birds is low; after three or four years reproduction recovers to normal levels or even higher because of the growth of new vegetation. Adapting to changed environments is critical to survival, but not every species can respond quickly. Populations of woodpeckers, owls, nuthatches, and others that depend on tree cavities for protection and nesting recover more slowly.

Weather, Foraging, and Food Supply

The time period at which birds migrate evolved as the most advantageous time to find an adequate food supply. So when an atypical cold and wet storm hit New England in the spring of 1974, it resulted in an unusual mortality of insectivorous birds, especially Scarlet Tanagers, which normally prey on large, heavy-bodied insects. The population of Scarlet Tanagers decreased by 33 percent the following year and 67 percent the year after that as reproduction was minimal and some adult birds died.

Birdwatching on a cold, brisk, windy day is usually a challenge. In a forested environment, birds will move away from the edge and into the denser forest. Woodpeckers shift from foraging on small branches to larger branches or the trunks of trees. Nuthatches stay on the trunk but birds like chickadees and titmice forage closer to the ground, where it is less windy, and dive into shrubs with each gust of wintry air. In more open areas birds may try to forage in the warming sunlight. Ospreys feed less in windy weather because the roiled water surface interferes with seeing potential prey. Smaller birds like the Tricolored and Little Blue Herons and Snowy Egrets may suspend foraging activity for as long as three days in severe weather, losing 6–12 percent of their body weight. Great Blue Herons, and probably other herons, feed more frequently when the sky is overcast than when it is sunny or raining. Presumably their prey of aquatic animals is more active on cloudy days or less able to spot the predator standing above them. Or the birds simply need more food.

Predators complicate the challenges posed by the vagaries of the weather. In a winter study of Common Redshanks, Eurasian shorebirds, the birds could choose to feed in one of two places: a productive salt marsh or less-productive intertidal flats. The colder the temperature, the more the birds preferred the salt marsh, but that habitat attracted predators like the Eurasian Sparrowhawk. At lower temperatures, sparrowhawks were more successful at catching their redshank prey because the redshanks had to spend more time feeding, but the redshanks still chose the more hazardous but more productive salt marsh. Better to obtain sufficient food at the risk of being eaten than face sure starvation.

Perhaps 60 million people in the United States and 25 million in the United Kingdom feed birds in the winter. Many people who maintain bird feeders around their home are concerned that they not only attract seed-eating birds, but also predators, and they do. Cornell Laboratory of Ornithology's Feeder Watch program determined that of all the predation incidents around bird feeders, the Sharp-shinned Hawk was responsible for 35 percent, Cooper's Hawks 16 percent, and cats another 29 percent. Predation by the Red-tailed Hawk, American Kestrel, and Merlin combined only accounted for 12 percent. The data suggest that although home feeders expose small birds to some risk from predators, they are no more dangerous than the wild and may actually provide some safety, as more birds are present to warn of a predator's approach. And the additional food provided may lessen the amount of time a bird has to forage elsewhere, reducing its total exposure to predation. Besides the predation question, people are concerned that the birds will become dependent on the feeder food in the winter, and fear that if the feeders become empty when the owners go on holiday, the birds will starve. A study of Black-capped Chickadees in the northeastern United States found that they obtained only 21 percent of their daily winter calories from bird feeders. Another study involved withdrawing food from feeders in areas where birds have been fed for 25 years and compared the survival rates of birds in those areas to birds that had never received supplementary feed. No difference. So go on a holiday and don't worry about feeding your birds.

Weather and Disease

Birds can contract about 60 diseases. Some are mild, others devastating. Birds may or may not develop an immunity and different species may respond differently to infection. The causes and effects of diseases are extremely varied and not everyday occurrences so birds often do not have the mechanisms to survive them. Avian deaths from disease may be closely linked to weather conditions, but most epizootics (diseases in animal populations) among birds tend to be localized, and are most frequent in birds that winter together in close proximity, such as waterfowl. Here are some of the more common and well-known avian illnesses:

AVIAN CHOLERA, caused by a bacterium, affects about 150 bird species, but is especially deadly to waterfowl in their wintering areas where the birds congregate and the bacteria are disseminated by bird-to-bird contact or ingestion of contaminated food. The bacteria also collect on the water's surface and spread into the air when the birds fly off. In the late 1970s cholera caused high mortality in Common Eiders wintering in open areas in the ice off the Netherlands and in the year 2000, 10,000 Baikal Teal died in Korea of the disease. Cholera outbreaks can happen anytime but are most common during the winter when cold conditions make waterfowl more susceptible to the disease as well as reducing the number of ice-free areas, causing the birds to crowd together. Fog and rain discourage birds from flying, further encouraging bird-to-bird contact. In March of 1994, an outbreak of avian cholera in severe cold weather killed an estimated 100,000 waterfowl in the Chesapeake Bay region. Cold temperatures along with a low-pressure center prevented the birds from dispersing northward. Earlier outbreaks in 1970 and 1978 killed up to 100,000 waterfowl each time. Avian cholera is detected virtually every year in waterfowl of the Mississippi Flyway. Researchers there do not think that cold weather initiated the epizootic but determined that colder temperatures were responsible for a higher mortality of infected birds. Warm weather can also exacerbate cholera epidemics. In 2012 the reduced snowmelt from a dry year was insufficient to freshen the water in the Klamath Basin of Oregon and the disease killed an estimated 10,000 waterfowl. The continuing drought in California is forcing more birds into smaller and less abundant bodies of water so the potential for future severe cholera epidemics looms.

AVIAN BOTULISM, caused by ingesting the toxin produced by the bacterium *Clostridium botulinum*, is responsible for hundreds of thousands of waterfowl, shorebird, and gull deaths in the western United States each year. An episode of the disease in Russia in 1981 killed more than a million birds. A study of the causes of mortality of 600,000-plus near shore and freshwater aquatic birds of 153 species over a period of 30 years in the United States indicated that botulism was the leading cause of death. The toxin acts upon the nervous system and kills birds by interfering with nerve impulse transmission. As paralysis sets in, birds can no longer fly or even hold their head up and often drown. The epidemics tend to be greatest in the late summer and early fall, especially with high ambient temperatures. Warm weather and heavy rain may precipitate botulism epidemics, but predicting outbreaks is difficult.

WEST NILE VIRUS (WNV) is a fairly new disease on the bird scene in the United States. Since first being found in New York in 1999, it has spread to all 48 contiguous states, infecting at least 300 species of birds with varying degrees of susceptibility. During the years 2001–2005, warmer weather, higher humidity, and increased precipitation contributed to a 30–80 percent higher incidence of WNV. Mosquitoes spread the virus to birds (and occasionally humans and other vertebrates). Most birds recover after about a week and become immune to another infection although crows and jays suffer a near 100 percent mortality. The disease is usually mild and rarely fatal in humans, with a mortality rate of about one in 1000. WNV is different from many other mosquito-borne illnesses because an urban-dwelling mosquito (*Culex pipiens*) is the primary transmitter. As the climate changes and we experience warmer winters and hotter summers, more mosquito eggs and larvae survive. In the spring, the few puddles or ponds that are available are saturated with nutrients that support mosquitoes. The adult mosquitoes emerge in large numbers because water levels are lower and there are fewer mosquito predators like frogs, lizards, and damselflies.

All environments undergo changes, both long and short term. Birds can adapt to these changes over evolutionary time. The soil, the vegetation, the topography, food supply, and potential nesting sites just don't change much year to year under natural conditions. The weather pattern changes from year to year and day to day so it is more difficult for birds to develop adaptations to deal with this moving target. But weather in any particular geographic area has parameters and limitations and it's the continuum of weather conditions birds have evolved to deal with. Climate change, global warming, however you define the increasing heat of the planet, has already modified some bird habitats and behavior and we will no doubt see more alterations. A disease, weather-caused or exacerbated by weather conditions, is one of those things that birds seem to have few adaptations to survive. It is often just a matter of bad luck. The same is true for many sources of human-caused bird mortality, which we will get to later in the book.

BIRD COMMUNITIES

How Birds Live Together

Birds . . . are sensitive indicators of the environment, a sort of "ecological litmus paper," . . . . The observation and recording of bird populations over time lead inevitably to environmental awareness and can signal impending changes.

—ROGER TORY PETERSON,
Peterson Field Guide to Birds of North America

Like many kids, I grew up fascinated by animals—live, stuffed, or fossilized. Catching snakes, frogs, and lightning bugs was part of it. At Chicago's Field Museum of Natural History, I loved seeing the dioramas of stuffed gorillas and ostriches with faux backgrounds of rocks and plants. And of course *Brontosaurus*, emerging from a swamp with a mouthful of soggy plants. I didn't really notice the plants—the trees, the ferns, the grasses—they were just there for decoration. Even in graduate school I would facetiously observe that the function of plants was just to give birds somewhere to perch. Slowly I realized that I was missing a great deal, as green plants are the basis of any ecosystem. Later, in my professorial role, I taught field biology to budding elementary school teachers. I encouraged students to look at a forest or grassland or lake and see not just a field of green or brown or an expanse of water, but a superorganism of intertwined life forms, all striving to survive in an environment of continual challenges. Birds are part of this large and intricate biological system, the framework of which is vegetation.

Birds live in a habitat of complex physical features and share space with many other organisms, but that's often difficult to picture based upon popular literature. Folktales and legends often describe one kind of bird, arising from the ashes, turning into a beautiful swan, or learning to sing. Kids' books include *Flora the Flamingo*, *Hoot Owl*, and *Redbird*. With a focus on the life of one bird or species, often missing is the larger context in which birds live and confront the challenges of survival. Birds do not live alone, ignore birds of different species, or avoid interacting with other animals. They don't go wherever they want and do whatever they wish. They are part of amazingly vibrant communities, which constantly pose opportunities and threats.

For many years, ornithological observation meant sitting and watching a bird and describing its behavior. Arthur Cleveland Bent, a businessman who became interested in birds as a child, spent nearly 50 years compiling 21 volumes of *Life Histories of North American Birds*. A typical excerpt reads: "When we think of the kingbird, even if it be winter here in the north, and he is for the time thousands of miles away in the Tropics, we picture him as we see him in summer, perched on the topmost limb of an apple tree, erect in his full-dress suit—white tie, shirt-front, and waistcoat." No criticism of Bent—his works contained enormous amounts of credible information—but a flowery description of one bird tells us little about its community. It wasn't until the middle of the 20th century that ornithologists began looking at assemblages of birds and seeing how their survival depended on interactions with their neighborhood and neighbors.

Every geographic area, large or small, is defined by a unique assortment of organisms surrounded and constrained by a distinctive set of physical elements: soil, weather, water, topography, and geology. This is an ecosystem—a collection of organisms interacting with each other and the physical environ ment—each one possessing its own physical, physiological, and behavioral attributes. Alexander von Humboldt, Prussian naturalist and explorer, wrote at the turn of the 18th century of the extravagant sounds and sights of the tropical forest and speculated that it was so dense "that there was simply no room to add another plant." Early naturalists wrote about flashing specks of light filtering through the thick canopy onto the forest floor, oppressive humidity, hordes of undulating army ants marching over thin soil, magnificent butterflies, colossal buttress roots supporting trees reaching to the sky, palm trees with thorny trunks, and masses of intertwined vines shooting upward and drilling downward. They wrote poetically about birds of every color, shape, and voice. The tropical forest that awed them is filled with squawking parrots in the canopy, antbirds shuffling the litter, brown creepers inching up tree trunks, toucan bills jutting from tree cavities, and hummingbirds flitting around dazzling flowers. Natural selection made these birds part of the tropical forest; few of them could survive elsewhere. The tropical forest and every other habitat have their own defined set of birds—the avifauna. The individual members of the avifauna do not act alone and unfettered; they are evolutionarily obliged to confine themselves to certain roles to survive.

The Arctic tundra, the Namib Desert, Lake Baikal, Central Park, and your backyard have much in common. They contain a set of plants and animals living in a framework of physical factors. A group of living organisms in

Tropical rainforest.

any area constitutes a community; it may be an insect community, a plant community, or a community of birds. Communities are not random assemblages—they evolved over long periods of time as each species in the community fit itself into the mosaic. In a human context, a community is a group of people living in a particular location, conveying the various roles that individuals play such as shopkeepers, teachers, and doctors. In the beginning a community might be rural with few people and the only doctor a general practitioner. As the community grows into a town and becomes more complex there may be pediatricians, surgeons, ophthalmologists, and family doctors. Avian communities and ecosystems develop similarly.

SUCCESSION TO EQUILIBRIUM: ARRIVE, THRIVE, AND DISAPPEAR

The earth came into being about 4.6 billion years ago and the first organisms appeared a billion years later. Birds didn't arrive on the scene until much later, more than 150 million years ago. Along the way, the earth underwent massive alterations in its geology, hydrology, and atmosphere. Organisms were simultaneously evolving and making their own contribution to the composition of the earth. Bird species transformed over time, existing ones continually replaced by new ones with improved adaptations for survival in the changing environment. There were once 200 pound penguins in Australia and 1000 pound Elephant Birds in Madagascar that stood 10 feet tall. These birds disappeared and others came to be. The survival time on earth of any species is ephemeral when measured in geologic time; the average life

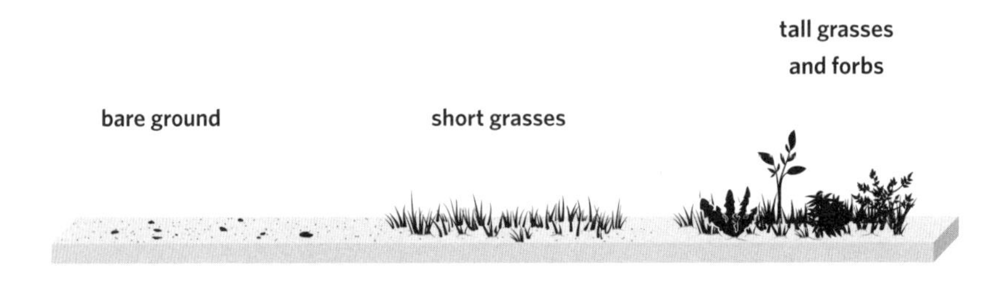

span of a bird species is about 125,000 years from appearance to extinction. Perhaps 160,000 species of birds have at some time lived on the earth, 16 times as many as the 10,000 species that exist today. So bird communities continually change, as some members are added and others drop out.

Geological time is a long perspective, but ecosystems develop on a much shorter scale. As a kid in Chicago, I watched the empty lot next to our house, denuded by construction activities, transform into an ecosystem. This abandoned plot of dirt was invaded by plants we call weeds. Grasses arrived, multiplied, and were eventually replaced by shrubs and trees. Insects appeared, followed by rodents, snakes, and birds. Had it not been again denuded by new construction it would have become a deciduous forest. This procession happens everywhere all the time. Fires, hurricanes, floods, earthquakes, and human disturbances yield a blank canvas and raw materials for new ecosystems. Lakes fill in to become marshes, grasslands, and perhaps forests. Islands arise from the sea, hot, rough, and naked, but are ultimately covered by plants and soil and inhabited by animals. These are examples of succession, the gradual and predictable development of communities. The process

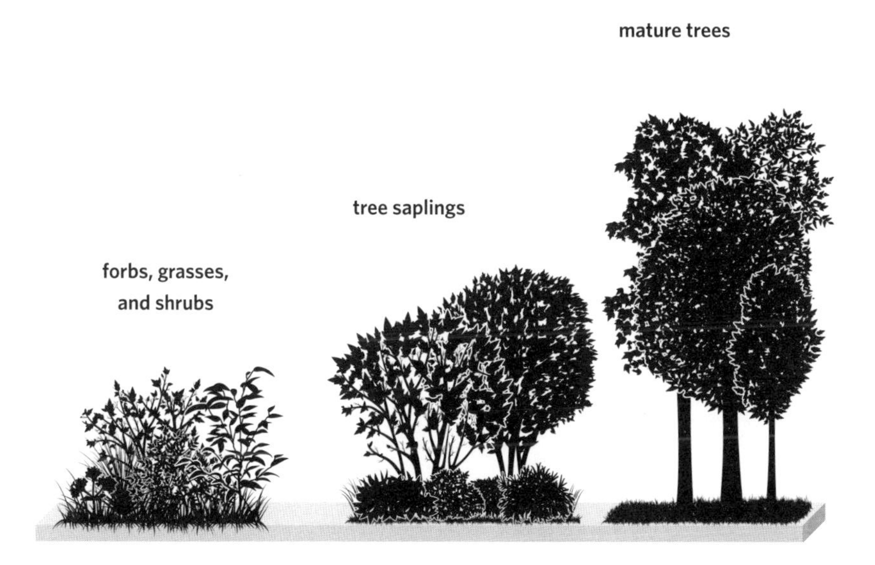

Destruction or severe disturbance of a forested area, partially or completely denuding it because of fire, flood, biohazard, or other human disturbance, leads to eventual recovery of the habitat by successional stages.

works the same way in every climate. In northern temperate zones, hardy low-growing grasses and forbs begin to colonize the area. Eventually, hardy shrubs take hold and ultimately loom over the grasses and forbs, shading and usurping their sunlight and nutrients. Tree seeds carried by the wind, water, or birds arrive, germinate, and become saplings. The trees in turn shade and outcompete the shrubs and form a young forest. The saplings become tall trees, living for many years, and the forest's composition and structure remain essentially the same for centuries.

Changes in bird species composition parallel the changes in plant species. Grassland birds are replaced by birds of the shrubs, followed by birds of the forest. The more complex the plant community, the greater the number of different bird species and the more individuals of each species. Just as we can predict the end result of plant community succession in any specific geographic location, we can predict the succession and ultimate constitution of avian communities.

Succession never ends but it slows down considerably as an ecosystem can hold just so many species of birds (or plants or insects), and every new species has the potential to cause the extinction of an existing one. New species arrive at all stages, although not all of them survive. In fact, most don't. A well-studied example is Krakatoa, a volcanic island in Indonesia lying between Sumatra and Java. Krakatoa erupted with explosive force in 1883, killing 36,000 people and destroying two-thirds of the land. The eruption was so powerful that tsunami waves rocked ships off the coast of South Africa. Although a tragedy for the local populace, it afforded opportunities for gathering firsthand knowledge of the development of an ecosystem and its avifauna. In 1889, six years after the eruption, seeds had blown or washed in and vegetation started to recover, but there were no resident birds. By 1908, more plants had appeared and 13 species of birds had taken up tenancy. By 1924 tropical forest plants were abundant and 28 bird species were breeding there, although two previous occupants had disappeared. In 1934, 171 plant species were identified along with 29 bird species, but three earlier bird species were gone. In 1952 there were 33 bird species, but three former species had disappeared. And in 1984–1986, there were 36 resident bird species, but four previous occupant species were absent. Today, 38 species

This 1888 lithograph shows the Island of Krakatoa as it might have appeared when it erupted in 1883.

survive. Just like the plants, the bird species arrive, thrive, and disappear as succession occurs until equilibrium is reached.

Recovery from geologic events can differ greatly. The volcanic island of Surtsey, 20 miles south of Iceland, arose anew from the Atlantic Ocean during the years 1963–1967; today only 12 species of birds, mostly seabirds, survive on the isolated island. Gulls are the most abundant birds on the island and have significantly influenced the growth of plants on the island because they fertilize the soil with guano. Eventually the avifauna will expand as the vegetation flourishes. The eruption of Mount St. Helens in Washington State in 1980 covered the surrounding countryside in ash, devastating the entire avifauna of the enormous blast area, but four days after the eruption birds were seen flying over the site. Since then more than 80 bird species have colonized the mountain from the surrounding ecosystems. Details of succession vary, but the underlying concept is that ecosystems develop in an orderly and predictable way and at maturity are dynamic but stable. And in every ecosystem each bird species occupies a particular niche.

NICHES AND HABITATS: MAY THE BEST BIRD WIN

Each Nutch in a Nitch knows that some other Nutch
Would like to move into his Nitch very much.
—DR. SEUSS, *On Beyond Zebra*

Each bird species in an ecosystem occupies an ecological niche, defined as its relationships with the living and non-living portions of its environment. The niche includes all the variables a bird has to deal with to survive such as climate, food, competitors, predators, and vegetation structure. The niche can also be described as the bird's role in its community, or its job, and the habitat (the physical place it occupies) as the bird's place of employment. Some birds like jays and House Sparrows have wide-reaching niches because they are omnivorous and nest most anywhere. Other birds like hummingbirds, Ospreys, shorebirds, and pelicans have specific needs and thus narrow niches.

Pioneering ecologist and zoologist Joseph Grinnell was a potent influence in moving ornithology from a collecting and cataloging venture to one that examined the lifestyles and habitats of birds. In 1904 he accurately stated that the range of the Chestnut-backed Chickadee of the Pacific Northwest is due to "atmospheric humidity, with associated floral conditions." In a coniferous forest, warblers prefer tree branches while thrushes ply the ground. Spotted Sandpipers strut on the edge of a creek along the forest's edge while kingfishers perch furtively overhead. The narrow-winged Sharp-shinned Hawk nimbly flies through the forest while the broad-winged Red-tailed Hawk soars overhead. In the ocean, various seabirds forage at different depths and different distances from shore and nest on cliffs, on the ground, in burrows, or in trees.

Narrowly circumscribed niches allow multiple species to coexist in the same physical location by sharing resources. Consider a human community with retail businesses like a hardware store adjacent to a candy shop. They coexist because they are in separate niches and use different resources (customers with different needs). So woodpeckers, warblers, and sparrows coexist as they have fairly different niches, but what happens if more birds move in? Add more woodpeckers and sparrows and new birds like creepers and nuthatches and vireos: how will things work now?

Birds of one species may have the potential to utilize an assortment of foraging sites, roosting spots, and nesting locations, but with many bird species in an ecosystem, there is going to be some overlap in their requirements, and sharing can only go so far. The Russian biologist G. F. Gause developed the competitive exclusion principle which simply states that no two organisms can occupy the same niche if they have exactly the same requirements. How different they have to be depends on the environment and its resources. As we shall see, even seemingly subtle differences can allow coexistence.

In his 1859 *Origin of Species*, Darwin said, "We have reason to believe that species in a state of nature are limited in their ranges by the competition of other organic beings quite as much as, or more than by adaptation to particular climates." This competition for limited resources results in what Herbert Spencer called "survival of the fittest" after reading Darwin's original work. Darwin used the phrase later along with "natural selection," but they are not the same. The best fit birds are not just the ones who survive; they are the ones that go on to have the most young, perpetuating their genes. They are the ones who compete the best.

Interspecific Competition

The requirements of two different bird species may overlap just a little or not at all, like a vulture and cormorant, but when the niches of two species are so similar that they compete, we have interspecific (between species) competition. In the winter, Blue and Great Tits in Europe roost in tree cavities or nest boxes. Belgian researchers set up nest boxes with large entry holes that allowed both species of tits to enter and shelter during cold winter nights. But they did not provide enough boxes for all the birds. The larger Great Tits physically prevented the smaller Blue Tits from utilizing the boxes, resulting in a higher survival rate of Great Tits. The following spring the expected population increase occurred in the Great Tit but not in the Blue Tit population, showing that the size and aggressiveness of the former enhanced their survival.

In the forests of Finland we find the Willow Tit, Great Tit, and Crested Tit ("tit" comes from a Norwegian dialect word *titta*, meaning small). These small, mainly insectivorous birds supplement their diet with berries and seeds in the winter. In the woods, the Willow and Great Tit forage in the same tree: the Willow Tit uses the upper and outer branches and the Great Tit feeds in the lower and inner parts. But if the Crested Tit arrives, the Willow Tit moves to the lower and inner branches and the Great Tit moves away to trees on the forest edge. By altering their foraging behaviors to reduce competition, they all get food and survive the winter.

The classic example of similar birds sharing the habitat is MacArthur's warblers. Robert MacArthur was an influential ecologist and a founder

The Great Tit, a woodland bird, has easily adapted to human habitation.

of the field of evolutionary ecology. For his PhD thesis at Yale, he studied five species of warblers in the coniferous forests of the northeastern United States. The breeding ranges of the Cape May, Blackburnian, Bay-breasted, Yellow-rumped, and Black-throated Green Warblers overlap. Until MacArthur's study, ornithologists assumed that these birds looked and behaved so much alike that they might be exceptions to the competitive exclusion principle and actually share the same niche. MacArthur figuratively divided the trees into 16 different vertical and horizontal zones. He observed the foraging habits of the birds closely and found that if different species fed on the same coniferous tree, they foraged in different parts. This study of niche segregation is often quoted as a model example of how birds with similar niches can coexist.

Intraspecific Competition

More common than interspecific competition is intraspecific (within species) competition because birds of the same species have exactly the same needs. A study of Northern Gannet nesting colonies in the United Kingdom discovered that bigger colonies grew more slowly than smaller ones. Gannets from larger colonies faced greater competition for food and were obliged to fly further out to sea to forage than birds from smaller colonies. Longer foraging forays meant fewer trips, less provisioning for the young, and lower chick survival. Similarly, the population growth of Little Penguins on small islands off of eastern Australia slows as colonies get bigger since the adult birds have a short foraging range and feed at the surface or moderate depths, effectively limiting their food supply. As the colony grows and competition for food increases, the adult penguins spend more time foraging for scarcer food and the young get less food, resulting in an inverse relationship between the weight of the nestling penguins and the size of the colony.

The Black-bellied Seedcracker, a finch of the equatorial rainforest areas of Africa, feeds on two sedge plant species, one with hard seeds and one with soft. This bird species is divided into two morphs (morphological types), distinguished by width of their lower bill; one morph's bill is wide and the other's narrow. When both sedge seeds are abundant, the overlap in the morphs' diet of seeds is considerable. When seeds are scarce at the end of

MacArthur's warblers. Shaded portions of trees reflect those areas mostly used by the species indicated.

the dry season, the wide-billed morphs primarily choose the hard seeds and the narrow-billed birds eat the soft sedge seeds as well as other foods. The difference in bill width is genetic and the mating of a wide-billed bird and a narrow-billed one results in offspring that are either wide- or narrow-billed. Recently, a third morph with an even wider bill that fed on even harder seeds was discovered.

In rocky habitats of Eurasia we find the Western Rock Nuthatch of Croatia, Greece, and Turkey, and the Eastern Rock Nuthatch of eastern Asia. In the eastern and western extremes of their ranges, the species look very much alike and eat similar foods, but where the populations overlap, in Iran, the Asian bird has a larger bill than the European bird. When they share the same space, not only do they eat different-sized foods, but their eyestripes are different. Dissimilar bill sizes reduce competition for food; dissimilar eyestripes enable them to recognize individuals of both species, lessening confusion in figuring out who is who, reducing time and effort in territorial or courtship display.

The One-Third Difference Phenomenon

Pairs of different species sometimes look very similar. These "sibling species" are presumed to have been split off from one former species. The geographical ranges of Lesser and Greater Yellowlegs, North American shorebirds,

Where the ranges of Western and Eastern Rock Nuthatch overlap, the birds display differences in their eyestripe and bill size.

overlap considerably. The birds make their living by probing the muck of wetlands for invertebrates. The Lesser Yellowlegs bill is about as long as its head while the similar but taller Greater Yellowlegs has a bill at least one-third longer. The Greater Yellowlegs also eats frogs and crayfish and skims the water's surface in search of fish, which the Lesser Yellowlegs, eating smaller items, never does. Another example is the Cooper's and Sharp-shinned Hawks. The two birds share their looks and geography but the Cooper's Hawk is almost one-third larger. Cooper's Hawks feed mostly on medium-sized birds like robins and starlings, meaning that their average prey is more than double in size that of Sharp-shinned Hawks, which feed on smaller songbirds. In eastern India, four species of kingfishers all live in the same mangrove habitat, but differ in food preferences and behavior—the height of their feeding perch, distance covered on foraging forays, and the size of prey, all reflective of their different body sizes. The larger the bird, the bigger and higher the perch, the farther it flies to feed, and the larger the food items.

G. Evelyn Hutchinson, often considered the father of modern ecology, put forward an explanation in his classic 1959 paper ("Homage to Santa Rosalia or Why Are There So Many Kinds of Animals?") as to why such differences in size occur and why diversity is limited. Hutchinson proposed that there needs to be about a one-third difference in the sizes of animals with similar niches in order to coexist. This is sometimes called the Hutchinsonian ratio. Lesser and Greater Yellowlegs and the Sharp-shinned and Cooper's Hawks show that difference while in the case of the Indian kingfishers, the largest

Greater (left) and Lesser Yellowlegs (right).

is 1.25 times larger than the next largest bird which is 1.12 times larger than the third largest bird, which in turn is 1.5 times as large as the smallest bird. Other examples of similar species differing in size by one-third are the Lesser Spotted and Greater Spotted Woodpeckers, the Merlin and Peregrine Falcon, the Whimbrel and Eurasian Curlew, Snow and Ross's Geese, and American and Fish Crows. Although it has became a sort of rule of thumb that birds that have similar niches can only coexist if they are about one-third different from each other, the idea is controversial. DNA studies of sibling species reveal that some pairs are closely related and that some have greatly diverged genetically.

Territories: Get Outta My Way

Anywhere a bird goes is its home range and within that may be a defended area: the territory. A territory serves to spread out competing individuals into their own spaces and reduce competition. Robert Ardrey in *The Territorial Imperative*, argues that territoriality is innate in all animals, including humans, and entitles his first chapter "Of Men and Mockingbirds," appropriate since the Northern Mockingbird is known for its strongly territorial behavior. The defended space could be a foraging area, an area around the nest, the nest itself, a place for courtship, or a roosting area. Territories are generally held only during the breeding season but some birds hold winter territories to protect their foraging sites.

Food is often a major factor in determining whether or not territories will be held and because food supplies fluctuate, territorial behaviors sometimes change. Hummingbirds are good examples as the concentration of nectar available among flowers changes often. One day a botanist colleague of mine and a few of his students trucked 500 containers of flowering plants to Yosemite National Park. They placed the plants on the ground in a predetermined grid and out of sight of any other blooming flowers. Within five minutes, Anna's Hummingbirds found the flowers and established territories around several of them. A few hours later the territorial boundaries changed as the nectar concentrations of the flowers changed.

Rufous and Calliope Hummingbirds in Nevada feed mainly on one species of flower. The Rufous Hummingbird tends to feed at heights of 8 inches or more above the ground and defends its food source vigorously. The smaller and faster-flying Calliope Hummingbirds do not hold territories. Instead, they simply raid the territories of the Rufous Hummingbirds by zooming in below the Rufous Hummingbird feeding zones, which change as flowers bloom and wilt. The smaller size and higher metabolic rate of the Calliopes make territoriality too energetically expensive so they simply robbed Rufous Hummingbird territories but avoid most confrontations by feeding closer to the ground.

Several species of closely related flycatchers in western North America—the Dusky, Gray, Willow, Alder, Pacific-Slope, Hammond's, and a few more—are notoriously difficult to tell apart as they look and act very much alike. Like MacArthur's warblers it appears that they would strongly compete for food and nest sites during the breeding season. But the birds, instead of feeding in different parts of a tree or squabbling over nest sites, hold both interspecific and intraspecific territories to spread themselves out and reduce competition.

Red-winged Blackbirds inhabit marshes, wetlands along roadsides, and golf course ponds where males establish territories, sitting on top of cattails, singing and flashing their red shoulders to attract females and defending against intruding males. (One experimenter captured territorial males, painted their red epaulets black, and released the birds. Almost immediately they were attacked by red-shouldered males and driven off.) After males fill

Red-winged Blackbird displaying epaulets in territorial defense.

the habitat with territories there are still floaters on the edges—males that were unable to establish a territory but are ready to fill the position of a male that leaves or dies. There seems to be little difference in the fitness of the floater males compared to the territory holders because once a floater is able to obtain his own territory, he does just fine in defending it. Territory holders appear to have more at stake and defend the territory vigorously, whereas a floater can abandon a challenge to a territory holder with little risk.

FORAGING GUILDS AND DIVERSITY

Studying bird communities is a difficult business partly because some bird species are more numerous or obvious whereas others are harder to find. Bird community composition also changes with the seasons, further complicating any analysis. So we need to satisfy ourselves with a snapshot that reflects an avian community at one particular time.

Because bird bills define much of a bird's niche, the study of foraging habits has become a major tool in the study of bird communities. One study

Pileated Woodpecker (sometimes mistaken for the similar but extinct Ivory-billed) represents the woodpecker guild, with about 200 species nearly worldwide that make their living in very similar ways.

of 22 species of insectivorous birds in a deciduous forest in New Hampshire—warblers, thrushes, vireos, chickadees, sapsuckers, and wrens, among others—classified their feeding maneuvers into 17 different categories. Researchers noted behaviors like hawking, probing, and sallying, the height at which these maneuvers occurred, and in which of eight species of trees they took place. Based upon feeding habits, the 22 bird species were grouped into guilds—assemblages of birds that feed in similar ways. ("Guild," in Medieval times, meant a group of craftsmen, workers, or merchants who shared the same interests.) Avian guilds were defined in this case as ground foragers, tree trunk and branch foragers, canopy feeders, and those that feed in other parts of the vegetation. Within each guild the birds were subdivided by their differential use of foraging substrates (such as bottom or top of leaf), the use of different tree species (such as oak or maple) and foraging maneuvers (such as hovering or probing). This research, like many other similar studies, demonstrates that avian communities can be defined and studied by using categories of foraging styles. We can also define guilds by taxonomic relationships like woodpecker guilds or habitat relationships like shorebird guilds, or even guilds like seedeaters that include not only birds, but rodents, insects, and others.

Foraging guilds give us both a flowchart and blueprint of an avian community, so let's look at the guilds found in the typical ecosystem and how they fit into the avifaunal community. As we move along, consider what would happen if a guild of birds were to disappear.

DETRITIVORES OR DECOMPOSERS are the vultures, obligate scavengers that recycle dead bodies and limit the spread of diseases. They fly and soar slowly over wide areas, often in groups, to increase their chances of discovering food. They are big because the unpredictability of their food sources necessitates survival on their body reserves between feeding bouts. Condors can consume more than 4 pounds of carrion at one feeding and Turkey Vultures can put on enough fat to survive for two weeks without eating. Vultures don't generally pose much of a competitive threat to other bird species as most vultures specialize in consuming carcasses. Some facultative scavengers like crows and ravens might share in the meal if they have the opportunity, and in dense environments like a tropical forest where it is difficult for vultures to maneuver, crows, ravens, and jays might become the major scavengers. You would think that putrefying corpses would be the major attraction, but vultures prefer fresh kills. A vulture's major competitor is bacteria, which rapidly make the dead body unpalatable.

GRANIVORES account for about 15 percent of the avifauna, mainly finches and sparrows, in a deciduous forest; in a coniferous forest they comprise about 35 percent of the bird species; in grasslands 60 percent; and in grain fields up to 90 percent. Granivores may consume a large portion of the seed production of an ecosystem, up to 20 percent of the annual seed production of a coniferous forest. Seeds have to survive on their built-in resources until they germinate, so they are packed with nutrients—up to 65 percent carbohydrates, some fiber, some protein, and a bit of fat. Granivores cannot survive on seeds alone, however, and need to supplement their diet with insects for more protein.

HERBIVORES eat approximately 10 percent of the plants (roots, shoots, or leaves) in any ecosystem, but only about 3 percent of bird species use vegetation as a major source of food as plant parts are fibrous, only 20 percent digestible and contain less than 20 percent protein. To make the most of plant eating, herbivorous birds strategically select plant parts high in protein and low in fiber. The Vegetarian Finch of the Galapagos Islands feeds primarily on buds, leaves, flowers, fruit, and soft bark under twigs; the bird has a parrot-like bill, a large gizzard, and an exceptionally long intestine to digest plant materials. Like granivores, most herbivorous birds have to supplement their diet with insects for protein. One exception is the Plantcutters, small South American birds with beaks, jaws, and palates modified to macerate plant material, a muscular gizzard, and highly folded intestines that allow them to digest plants so thoroughly they have no need for animal protein.

FRUGIVORES comprise about 12 percent of all bird species. Half are songbirds, but a variety of other birds also consume fruit. High in carbohydrates and low in protein, fruit has a fat content varying from 1 to 67 percent. Many fruits have indigestible parts such as skin, seeds, or a hard seed coat, or contain distasteful or toxic compounds. Frugivores are mostly residents of the Neotropics and are major plant dispersers; many plants evolved fruits with characteristics specifically to attract dispersers. Seed-dispersing frugivores are especially important in succession—revegetating damaged ecosystems or developing habitats on new islands. In supplying this service to plants, frugivores ensure their own survival by providing a continual supply of food. There is little competition or specialization among frugivores because fruits tend to be superabundant when they are ripe and fruit eaters need to be able to handle whatever is available at the time. The dominant frugivores in tropical America are the various species of toucans, as caricatured by Toucan Sam, the mascot of Froot Loops cereal (which contains no fruit, just fruit flavoring).

INSECTIVORES, birds that survive on arthropods, account for approximately 60 percent of the world's birds. Insects are mainly protein and very digestible, so it is no wonder that more than 7400 species of birds feed on invertebrates, including 44 mostly insectivorous hawks. There is such a variety of exploitable arthropods that birds employ a wide array of foraging behaviors like hawking, sallying, gleaning, or probing. The Blue-tailed Bee Eater is a dedicated insectivore specializing in bees and wasps, which it beats on a branch to kill and soften for easier swallowing. The tropics offer bugs all year but winters in temperate zones make arthropods scarce so permanent residents have to be flexible and find dormant insects, larvae, or eggs, switch to another food source, or leave. Downy Woodpeckers probe the crevices of tree trunks, galls, the stems of weeds for arthropod larvae, as well as seeds and berries. Chickadees will eat the seeds of coniferous trees, berries, and small nuts. They have even been observed picking the fat off of dead squirrels. Beechnuts are an important winter food for the Great Tit. Both chickadees and tits also cache considerable amounts of food, as much as 100,000 items in a season.

Flycatchers, warblers, swallows, and swifts are dependent on active insects so in the winter they migrate to the tropics where they have access to the creatures. What happens to the resident birds when all these insect-eating migrants arrive, sometimes doubling the bird population? Insectivorous birds that are permanent residents in the tropics tend to be specialists, surviving in narrow foraging niches. For example, 11 percent of insectivorous birds in the upper Amazon basin feed solely by acrobatically gleaning insects off aerial leaf litter (dead leaves hanging from understory plants), as some antthrushes and ovenbirds do. Although dead leaves are less numerous than live ones,

Blue-tailed Bee Eater.

dead leaves hold more arthropods and thus offer a higher energy yield. Migratory birds arriving in the tropics are opportunists and survive by feeding wherever and however they can. An exception is the Worm-eating Warbler from the eastern United States. In the spring it spends about 75 percent of its time searching live leaves, but when wintering in Central America it forages 75 percent of the time on dead leaves, like tropical ovenbirds and antthrushes.

Insectivorous birds are important for keeping insect levels under control in forests and reducing plant damage. Researchers in southern Sweden used nets to exclude birds from tree trunks and branches. After four weeks, plant-eating arthropods on the covered tree parts increased by 20 percent. A similar study in a Jamaica coffee plantation resulted in a 60–70 percent increase in arthropod populations on the trees. In 1918 the State of Michigan estimated the worth of insectivorous birds to farmers at $10,000,000; pretty amazing considering that the state at the same time considered the total worth of all its game mammals, birds, and fish at $500,000. Insectivores do not have as great an influence on insect control in temperate ecosystems as they do in tropical ones because cold winters keep insect populations dampened.

NECTARIVORES are small but important pollinators such as hummingbirds, sunbirds, honeycreepers, and honeyeaters. More than 900 birds feed on nectar and serve as pollinators to more than 500 plant species, pollinating as much as 10 percent of wild plants in the tropics and perhaps 6 percent of agricultural crops such as bananas and papayas. Pollinators are especially important for isolated populations of plants because wind pollination is unreliable. As long as the avian population stays healthy, bird pollination works well, but consider the New Zealand gloxinia, a flowering shrub that is largely dependent on the Bellbird and Stitchbird for pollination. When the birds went extinct on New Zealand's North Island in the 1870s, the plants became far less productive than they once were.

CARNIVORES, which include the hawks, eagles, owls, falcons, caracaras, and relatives, are at the top of the food chain. The vast majority (90 percent) of all raptors live either exclusively or mainly in the tropics, a reflection of the productivity of the tropical ecosystem. Some are specialists like the Osprey, the only diurnal bird of prey that feeds exclusively on fish, the Black-chested Snake Eagle of Africa, Pel's Fishing Owl, and the bird-eating Peregrine Falcon. But most raptors are generalists and opportunists and have few predators so they are mainly limited by competition for food. Of the nearly 500 raptorial birds, about one-third are nocturnal, lessening the competition for prey.

Bird Species Diversity

Within an ecosystem, avian communities of various foraging guilds all make a living—survive—by competing for food while simultaneously avoiding *being* food. Although all ecosystems and avian communities work in the same general way, each one is unique. So how can we compare them? One of the most important goals in describing an avian community is to establish a baseline against which any future changes can be measured. What are the possible impacts of projects like wind turbines, dam construction, traffic, or increased noise, for example? If we know how the bird community works in a particular system, we can measure the potential effect of these projects and suggest possible mitigation measures. So how do we assess the avian environment? One of the most potent ways to evaluate a bird community and its changes over time is to measure diversity.

Ecologists have for years argued about what diversity means and how important it is. A workable definition is that diversity is a combination of species richness and equitability, a combination of the number of bird species and the numbers of individuals in each species. It's obvious that a community of three species with 10 individuals in each is more diverse than a three-species community with 26 individuals in one species and two in each of the others. But intuition goes just so far.

It is important to have some quantitative measure of bird diversity in a habitat that we can use for comparison from year to year. If the numbers change—fewer birds of one species surviving, perhaps—something is happening that needs to be investigated. Perhaps more importantly, the measurement of bird populations provides an accurate reflection of the health of the entire ecosystem.

A classic paper by Robert and John MacArthur proffers an elegant solution to comparing the diversities of birds in a habitat. After several weeks in the field, we can get a good handle on the number of bird species and the number of individuals of each species in a particular area. Using the MacArthurs' simple mathematical formula we can compare the bird diversity of different habitats. We know that birds segregate themselves at least partially by their foraging locations, so it's essential to measure the physical structure of the habitat—the vegetation. We do this by determining the density of the vegetation at different heights. Think about walking through the woods, having to step over the underbrush and walk around trees and shrubs, encountering open spaces and impenetrable brambles. Shrubs at 2 feet have a different configuration than a tall coniferous tree at 10 feet and both differ from a dead tree. Clearly, the structure and abundance of the plants are different in each habitat and the birds react to those. The idea here is that there is a close relationship between the plant structure of a habitat and the bird species diversity. Complex vegetative structure means more niches, although scientists have argued the nuances of this concept ad infinitum.

Is diversity important? Ecologists and conservationists have contended for years that diversity means stability, and that complex ecosystems are more likely to resist and recover from perturbations than simple ones. Not everyone agrees, but the idea does make some sense. Consider a mechanical pocket watch. The watch has a bunch of parts, some more essential than others. Remove a few parts like the crystal, second hand, and numbers on the watch face, and the instrument will still work. But if you continue removing parts at some point the watch will malfunction. Do the same with an ecosystem. Remove one of the frugivores and other frugivores will likely fill in and provide seed dispersal. Eliminate a nectar-supping pollinator and another one will probably take its place. Kill off one species of hawk and the forest will probably continue to function more or less as before. But continue to simplify the system by removing more birds and the ecosystem will eventually suffer deleterious effects.

If a guild of birds disappeared there would be major changes in both plant and animal communities, or even geological effects. The Bay of Fundy, touching New Brunswick, Nova Scotia, and the state of Maine, is frequented by several species of migratory shorebirds. The shorebirds feed on amphipods (small shrimp-like creatures); one sandpiper might eat as many as 10,000 per day. The amphipods feed on diatoms—tiny, mostly one-celled algae. Although miniscule, the diatoms are extremely abundant and play a significant role in the tidal ecosystem, producing adhesive chemicals that help stabilize shoreline sediments. If the shorebirds were to disappear, amphipods would increase, diatom numbers would decrease, and the shoreline would erode. That's a pretty significant outcome. Another example: the Eurasian Jay is a major disperser of many European oak species. The jays pick acorns from trees and bury them in abandoned croplands, pastures, and openings in the forest, and retrieve them in the fall when food is scarce. A pair of jays can scatter and hoard thousands of acorns in a season. Although the birds retrieve a good percentage of them, many acorns go uneaten and become oaks. Any decline in the jay's population would reduce the distribution and survival of many oak species.

PREDATION

Predation can affect a bird community directly by reducing prey numbers or indirectly by causing birds to change their behavior. Some researchers argue that the non-lethal effects of predation (not being eaten) are at least as strong, if not stronger, than the lethal (being eaten) effects. If birds change their behavior to avoid predation and spend less time foraging, they or their nestling young may starve. The threat of predation may keep birds from defending their territory or attracting a mate. Raptors are a major force in creating and maintaining the structure of the avian community by keeping the numbers of some birds down and sometimes reducing the overexploitation of a resource.

Hawks and owls, for example, eat small birds, but also cause those same birds to forage less frequently for fear of being eaten. One study in Finland measured the distances of songbird nests from a nesting European Kestrel; as expected, there were fewer nests of songbirds near the kestrel nest than there were farther away. Kestrels prefer open habitats and prey species tend to space themselves a good distance from the predator's nest as they are more likely to see the approach of a predator and have the opportunity to escape. A similar study on the European Sparrowhawk, a forest-dwelling raptor, showed a minimal effect on the nesting distance of nearby songbirds because in this study there were more locations (bushes, shrubs, trees) to build nests in and more places for prey to conceal themselves nearby.

The mixed deciduous forest Wytham Woods at Oxford University has provided a venue for considerable bird research for many years. Researchers have studied the Blue and Great Tit populations there since 1947 and virtually every individual bird of these two species has been banded. Approximately 1000 nest boxes have been provided for the birds and careful observation has provided good data on all aspects of their lives, including predation pressure. Sparrowhawks take 20–25 percent of the tit population each year, but the tit breeding population remains steady. Predation makes room for immigrant tits, mostly young birds looking for an opportunity to mate and a place to nest. There's an ecological balance here—the predators eat some tits, which are then replaced by others. The number of available tits limits the predator population.

Intimidation by hawks can be put to practical use. In order to control Rock Doves in Trafalgar Square, the City of London paid a falconer to utilize trained Harris's Hawks to scare the birds. The pigeon population in 2005 was reduced from about 4000 to a few hundred. Hawks, falcons, and eagles are also used at airports around the world to scare birds such as gulls and blackbirds to avoid collisions with aircraft. In the United States alone more than 6000 aircraft-bird strikes are reported annually. Of course, raptors control mammals as well. One study estimated that a Barn Owl, in its lifetime, could eat 11,000 mice that would otherwise have consumed 13 tons of crops. The Lesser Kestrel prefers large insects like locusts that are often serious crop pests. Kestrel populations are declining across Europe, a study in Spain indicating the cause as the disappearance of grasslands and their replacement by sunflower fields (which makes hunting by the hovering kestrel more difficult).

Birds don't just casually sit around and wait to get picked off or cower in the presence of a predator. Prey birds employ numerous anti-predator

Lesser Kestrel (ranges from the Mediterranean to Asia) is experiencing population decline in Europe.

mechanisms to survive—escaping, hiding, possessing cryptic coloration, and feeding in flocks (more eyes mean a better warning system). They have also evolved specific behaviors like the broken wing act of the Killdeer, a distraction display to lure potential predators away from the nest. The bird fakes an injury to its wing by flopping and dragging it along the ground and the hungry predator follows. At a safe distance from the nest, the wing miraculously heals and the bird flies off. Other birds employ this ruse as well, such as the Mourning Dove and various Lapwings. (The name "lapwing" did not derive from this behavior but comes from the erratic flapping flight of the birds.)

Like zebras and tigers, many shorebirds have bands or stripes on their abdomen to break up their outline, making them harder to spot. Nighthawks, owls, and lots of sparrows and finches are various colors of brown and black to make them inconspicuous. The American Bittern's bold brown vertical neck stripes help it blend in with its marsh habitat of reeds and cattails. To narrow its body profile and remain hidden from its prey and predators, the bird points its bill skyward and slicks its feathers. On windy days, the bird sways slowly back and forth, like a bunch of reeds moving in the wind.

Sunbitterns of South and Central America inhabit open edges of rivers in tropical forests and are susceptible to predation by raptors. Not being fast fliers they prefer to walk unless crossing a river. When threatened, they spread their wings, tilt them vertically, and raise their tail to fill the space between. Their rufous-, gold-, and black-patterned wings resemble the menacing eyes of a large animal. Males, females, and young birds all show this pattern, with no intermediate plumages, indicating that the display is a defensive mechanism and not one of courtship.

One behavior that you are likely to have observed as it happens in the open and engaged in by many species is "mobbing"—ganging up on a perceived predator. When small birds detect a predator, they issue alarm calls and start to fly toward the predator in the hopes of chasing it off or at least annoying it. A few birds begin mobbing but eventually a number of them become involved. Gulls and terns are well known for their mobbing behavior but many songbirds use this tactic too. A predator sitting on a branch or limb or fence post, be it a cat, crow, hawk, owl, or some other creature, depends upon stealth to catch its prey and mobbing exposes them. If the predator is near a bird's nest, the mobbing might be particularly ferocious. Small birds have an advantage if they become aware of the threat early and become the aggressor themselves. Crows and soaring hawks are often chased by mobs of smaller birds; being less agile than the small pursuers the predators simply fly off. But agile fliers that prey on small birds such as Cooper's Hawks or Peregrine Falcons are sometimes mobbed as well. Turkey Vultures, Osprey, and even Great Blue Herons get mobbed even though they are little threat to songbirds simply because mobbing a big bird is an innate survival behavior. Birds most at risk, such as gulls and terns that nest in the open, are more likely to participate in mobbing whereas those birds that nest on ledges inaccessible to predators do not mob. In addition to mobbing the predator and sounding alarm calls, some birds defecate or regurgitate on the predator with amazing accuracy.

Some predators are greater threats than others so small birds sometimes use different calls to distinguish between the severity of the threats. Communally nesting Arabian Babblers issue two kinds of calls when they detect a predator. The oldest male babbler sits on the farthest outside edge of the

This 17th-century painting illustrates an owl being mobbed by a variety of songbirds.

colony, presumably because he is the most experienced at detecting threats. One call is a short, metallic-sounding "tzwick," his other a long trill. A cat, a snake, or other non-flying predator elicits trills, while a hawk generates mainly tzwicks with a more urgent meaning. Colonial nesting effects higher rates of survival and probably has evolved as a result of predation pressure since nesting in groups provides more eyes to detect and ward off danger.

EXOTIC SPECIES: INVADERS OR IMMIGRANTS?

I have visited nearly 100 countries and one of the first things I do in an exotic venue after deplaning is look for birds. Typically, the first bird I see is a non-native: it made a lasting impression on me when I visited South Africa for the first time, landing in Johannesburg, and there, on the tarmac at the foot of the airplane stairs, was a House Sparrow.

Exotic species have been around for some time. Until the introduction of the House Sparrow into the United States in 1861, the most common bird of farms, villages, and cities was the Chipping Sparrow, about which Audubon wrote: "Few birds are more common throughout the United States than this gentle and harmless little Bunting." In 1890 Eugene Schieffelin released some 60 European Starlings into New York City's Central Park. He was chairman of the American Acclimatization Society, a group of New Yorkers who were dedicated to introducing plants and animals from Europe. This was the time of the "melting pot" in the United States when immigration from Europe was in full swing and people wanted reminders of their old country. Schieffelin, a fan of Shakespeare, decided that every bird mentioned in Shakespeare's plays ought to be imported into the United States, such as the starling from Henry IV and the skylark from Romeo and Juliet. Several other species were introduced such as the Java Sparrow, Chaffinch, and European Robin, but unsuccessfully. The feral Rock Dove has been in North America since 1606, imported by the French at Port Royal, Nova Scotia, perhaps as food.

About the same time the European Starling arrived in New York, the Crested Mynah, native to Indonesia and China, was brought to Vancouver, British Columbia, and by 1920 an estimated 20,000 birds were living in the area. The Crested Mynah population stabilized as the birds never left the environs of the city because they couldn't tolerate the cold of the mountains. In 1950 the European Starling, spreading westward, reached Vancouver. The starling and the mynah had similar niches—they ate similar foods and both preferred to nest under the eaves of buildings. However, mynahs evolved in a warm location and starlings in a temperate one. Equipped with better insulation than the mynahs, the starlings could more successfully survive lower temperatures. In addition, even though the clutch sizes of the mynahs and starlings are both four to six eggs, because the mynahs come from a semitropical environment, their natural habit was to incubate their eggs only about half the day while starlings incubated all day and were more successful in raising young. So as the starling population increased, the mynah population decreased and by 2003 mynahs had disappeared from Vancouver.

An estimated 200 million starlings inhabit North America today. So what allows an exotic bird to survive in a new ecosystem? There have been thousands of other introductions of hundreds of species of birds across the world. Most don't make it or exist only in small numbers, but occasionally some flourish. A determining factor in the survival of a new arrival is the native range of that bird species: the greater the range, the more likely the species has a broad tolerance of climatic conditions and food choices. The European Starling's natural environment encompassed Europe and western Asia while that of the House Sparrow was even wider, spreading across eastern Asia, India, and northern Africa. Species with large ranges also tend to be reproductively prolific and spread quickly. Birds that are most likely to invade a new habitat are generalists and are likely to find the resources they need in a novel environment and are behaviorally flexible so they can adjust to a new niche. Diverse and undisturbed ecosystems are less prone to allow invading birds to establish a population in any numbers, but disturbed habitats leave open niches that may be accessible to exotics.

European Starlings and House Sparrows survived their importation quite well and spread across North America because they came from a similar temperate environment and invaded cities and agricultural areas inhabited by few other bird species. They are also comparatively bold. In the United States today 90-some species of free-living non-native birds subsist. Most survive in small, isolated populations, but others (besides the starling, pigeon and sparrow) such as the Ring-necked Pheasant and Mute Swan, are so abundant that most people do not realize they are imports. Most imports were deliberate, but about a third were accidental. In New Zealand, 130 non-native bird species have been introduced and 41 species have established populations at various degrees of survival. The European Blackbird and House Sparrow, introduced in the late 1800s to the islands by settlers to remind them of their former English home, were quite successful while the Cirl Bunting remains rare and the nightingale never made it.

When an invader arrives, it may quickly disappear, establish a small population, become peaceably incorporated into the ecosystem, become an agricultural or cultural pest, and/or overwhelm native populations of birds. There are 35 endangered native bird species on Hawaii, partly because of habitat destruction and partly thanks to alien bird species (as well as rats and mongooses). The islands have been colonized by 58 exotic bird species and another 82 introductions failed. The Japanese White-eye was introduced to the islands in 1929 and successfully invaded old-growth forests, usurping the niches of eight native bird species. The white-eyes impacted the food supplies of native birds, causing them to become stunted and more susceptible to disease. In a 19-year study on the island of Hawaii, researchers measured the body weights and bill and tarsi sizes of seven species of native birds and found that the body mass of the native birds decreased and their bills and tarsi became shorter, resulting in a lowered survival rate. The population of the Akepa, an endangered Hawaiian Honeycreeper, declined and finally crashed, parallel to the increase in the white-eye population.

The Eurasian Collared Dove is another "success" story for exotic species. Its original range in the 19th century was the warm temperate and subtropical zones of Asia and southern Turkey. In the early 1920s it began its spread across Europe. In 1970 they were brought to the Bahamas as pets, escapees reached Florida in 1982, and now they are found throughout much of North America, Europe, Asia, and even Iceland and the Arctic Circle. They tolerate a wide temperature regime and tend to nest around human habitation. Part of the reason for their quick spread is their reproductive capacity; even though they lay only two eggs, they have three to six broods a year. Eurasian

Collared Doves are so prolific that in Texas they are increasing at 15 percent a year; there, as in other states, there is an open hunting season on them.

Recent studies in New Zealand and Australia strongly indicate that exotic and native species distribute themselves along a gradient of habitat from natural or undisturbed to disturbed environments. Exotic and native birds separate themselves along a continuum ranging from plantation forests of exotic trees to undisturbed native forests, from thin and open vegetation to thick foliage, and shorter to taller vegetation structure. The more disturbed the environment, the more likely exotic species will establish a foothold. Generally, natural ecosystems are resistant to invasion by exotic birds, but when the ecosystem is disturbed, or carved into smaller parcels, invaders exploit the new conditions as new habitats provide open niches.

Native to Asia, the Ring-necked Pheasant was introduced into the states of Washington and Oregon in 1881 and 1882 and proliferated. Ten years later, the first pheasant-hunting season in Oregon yielded 50,000 birds. Croplands and adjacent brush are excellent habitat for pheasants and the birds quickly spread across the United States. Adaptable birds, they even populate tropical Hawaii. But today the pheasant population is declining in many areas. Clean farming, the use of pesticides, the disappearance of grasslands, grain fields converted to root crops, and changing weather patterns share the blame.

To sum up, avian communities are complex assemblages of interacting species. All these interactions are modified and affected by the structure of the vegetation and a myriad of environmental factors. Avian communities are organized to the extent that they are able to partition resources among the members so that they all survive to reproduce enough that they at least replace themselves. Birds have a variety of roles from scavenger to predator and nectarivore to granivore, and fit themselves into a niche, physically and biologically. Community organization is complex and it is often difficult to show cause and effect between one factor and another, but all the players interact at some level. Communities, like the one you live in, are in a constant state of flux, from hour to hour and decade to decade. All the organisms in the avian community have evolved survival skills to deal with all that turmoil. These assemblages of birds are the result of relentless competition over evolutionary time. As human society and our infrastructure grow we will continue to impact the natural communities of birds. Can they adapt? Perhaps. That is what the next chapter is about.

HUMAN INFLUENCES

What We Do To Birds

When not protected by law, by popular favor or superstition, or by other special circumstances, [birds] yield very readily to the influences of civilization, and, though the first operations of the settler are favorable to the increase of many species, the great extension of rural and of mechanical industry is, in a variety of ways, destructive even to tribes not directly warred upon by man.

—GEORGE PERKINS MARSH, *Man and Nature*

Alexander Wilson, in his *American Ornithology* of 1840, described the Passenger Pigeon as living "in such prodigious numbers, as almost to surpass belief; and which has no parallel among any of the other feathered tribes on earth." With a North American population of three to five billion birds, blackening the sky with its enormous flocks, the Passenger Pigeon was hunted with abandon. About 18,000 birds were killed every day by commercial harvesting in New York State in 1855. In a single year, a billion birds were killed in Michigan. The last bird died in a zoo in 1914. The Maoris hunted the flightless Moas of New Zealand to extinction by 1455. The last Dodo was seen in 1675, the species done in by hunting and the introduction of exotic animals to Mauritius. In the last century 100 birds have gone extinct, primarily by hunting and the introduction of predators (such as dogs, cats, rats, and snakes), and a few by habitat loss.

Imagine coming home one day and finding your house severely damaged, your bank accounts drained, and people you don't know hanging around. It would be intimidating and stressful. You might be able to adjust, rebuild, and recoup your physical and financial losses, perhaps move elsewhere if you could find suitable lodging, or you could end up homeless, destitute, and starving. Scenarios like this happen after major disasters, but humans are resourceful and most of us manage to recover. Now imagine you are a forest bird arriving at your breeding grounds after a long flight. Instead of an expected stand of trees you find a few scattered saplings, shrubs, and flowers interspersed between newly constructed streets and houses. Bird species you are not familiar with have invaded your old haunts. Your normal food supply is nowhere to be seen. Could you survive?

Evolution is a superb sculptor. Over hundreds of millions of years the machinery of natural selection has honed birds to pinnacles of near perfection, having discarded tens of thousands of species along the way that could not meet the challenges of the ever-changing earth. Varying climates, shifting continents, volcanic eruptions, fires, and other perturbations effected changes in organisms, purging some and refining others. Sometimes these changes are subtle and take many generations, but given sufficient time, ecosystems recover and within them communities of birds suited for survival in their particular habitats. But since the beginning of the industrial revolution the physical environment began to change at a much faster pace, leaving many birds behind. Very few places have not been seriously disturbed by human activity. Cities like Beijing and New Delhi are virtually devoid of avian life, and in metropolises like New York and London, House Sparrows and Rock Doves dominate the avifauna.

Even New Zealand, often envisioned as a pristine land, has seen 40 of its 115 endemic birds go extinct and 41 introduced species become residents. Botanist Joseph Banks, sailing with Captain Cook to New Zealand in 1769, wrote: "This morn I was awakd by the singing of the birds ashore from whence we are distant not a quarter of a mile, the numbers of them were certainly very great." New Zealand has lost more bird species than any other country and the majority of its land birds are threatened, endangered, or on the cusp of extinction. The only confirmed sighting in New Zealand of a Black-faced Monarch, common in eastern Australia, was in the late 1990s when it was caught by a cat, but it seems possible that the species once inhabited the islands. Every nation has some version of this problem. Some are addressing it by setting aside reserves, enacting environmental laws, and welcoming ecotourism, although many of those efforts are only moderate. Many other nations do little or nothing.

One interesting quirk in this dilemma is the Korean Demilitarized Zone (DMZ), which was created at the end of the Korean War in 1953 to minimize hostilities between North and South Korea. The DMZ is roughly 150 miles long and 2.5 miles wide. This strip of land has seen very little human activity and has become home to many species of plant and animals, including several thought to be extinct on the rest of the peninsula. The DMZ has become an important rest and refueling stop for birds on the Australasia-East Asia flyway such as the Red-crowned and White-naped Cranes and Siberian

The Black-faced Monarch.

herons. This unusual example shows the resilience of natural communities if they are left to their own ecological devices.

Without the interference of humans, the natural extinction rate of birds is about one per century. Over the last five centuries approximately 200 species of birds have gone extinct and today at least 1200 are in danger of disappearing. By around 2050 we will begin to see about one bird species per year go extinct. The environment and the particular ecosystems in which birds evolved and to which they are superbly adapted for survival are simply changing too fast for evolution to keep pace. Based upon the historical rates of the evolution of 540 vertebrate species, animals would have to evolve at a rate 10,000 times faster than they do now to adapt to climate change by the end of the century. And climate change is happening at a slower rate than other perturbations—the disappearance of habitats is probably the most urgent problem.

THE PERILS OF HABITAT DESTRUCTION AND DEGRADATION

Habitat destruction and degradation are greater immediate threats to birds than any other predicament. Turning grasslands into cornfields or replacing a forest with condominiums eliminates the vast majority of the local birds. Traveling across the Sacramento Delta and the San Francisco Bay by train, I get a good view of the productive estuary and wetlands: rafts of ducks, cormorants on pilings, an occasional pelican flying by, and bunches of shorebirds which have wintered here for millennia. Oil and sugar refineries, megastores, a racetrack, all manner of unknown abandoned factories, and one enormous sea of asphalt, which evidently had some past purpose, interrupt my view. The placement of some of these operations may have some logic, but why an auto repair garage had to be built on the edge of valuable waterfowl habitat is beyond me.

European researchers reviewed thousands of bird surveys conducted from 1980 to 2009 in 25 European countries and scrutinized the population trends of 144 bird species. They found a reduction of a staggering 421 million birds over that time period—and 90 percent of that number was caused by a decrease in some of the most common species such as Eurasian Skylarks, House Sparrows, Grey Partridges, and European Starlings. Not all species waned in number: Great and Blue Tits, European Blackbirds and Robins, Chiffchaff, Blackcap, and even some rarer species such as Marsh Harriers, White Storks, and Stone Curlews increased, likely because of conservation efforts and stronger legal protection. But the increases were small compared to the overall loss of birds. Expanding urbanization and farming methods such as the increased use of agrochemicals and the loss of hedgerows contributed greatly to these losses.

The reduction in bird numbers is generally proportional to the loss of suitable habitat up to about a 50 percent decline. So a 30 percent loss of habitat, for example, results in an approximate 30 percent reduction in birds. But when the loss of habitat exceeds 50 percent, entire species diminish disproportionately and may vanish entirely. Habitat fragmentation, breaking a large expanse of habitat into smaller parcels, usually with loss of vegetation between the tracts, changes the dynamics and constitution of avian communities. As a habitat is broken up, the borders of the patches become longer compared to the enclosed inner areas. This edge effect engenders physical differences such as increased wind, sunlight, and differences in food types and abundance. But the results are not always predictable. A forest in central Sweden fragmented by logging showed a decrease in the number of arthropods on the forest edges, but a study in Canada reflected an increase in bugs. Some studies indicated an increase in bird species diversity on the edges and others a reduction. The outcome of habitat fragmentation depends on how the environment was split up, what bird species are present, and what the adjacent habitats are.

Studies in a tropical forest of Ecuador showed that a reduction in the size of forest patches generally resulted in a decreased number of bird species, but the consequences of habitat disturbance varied. With low levels of disturbance that put the forest back to a recent successional state, the diversity of bird species increased because an increase in open space and sunlight prompted more plant growth and productivity; with higher levels of disturbance that sent the forest back to an early successional stage, the diversity of bird species decreased because few plants were available to support the avian

Deforestation in New Zealand, leaving a large forest edge.

community. Not all avian guilds changed in the same way. Nectarivore species declined as nectar-producing plants declined; frugivores survived after some habitat disturbance as berry bushes increased in abundance; granivores multiplied as habitat disruption resulted in more grassy plant species; and insectivores responded positively, perhaps because of the decline in nectarivores which competed for insects. A survey in Mexico after a major hurricane in 1988 showed declines in nectar and fruit eaters (because all fruits and flowers were knocked to the ground) versus only minor reductions in the numbers of omnivores and insectivores. Wintering bird species suffered less than residents because the residents tend to be specialists and the visitors are adapted to be generalists on their wintering grounds.

In the United States the composition of suburban bird communities depends largely on how the natural habitat was altered. In starkly urban areas with little natural habitat nearby, a few species such as House Sparrows and European Starlings appear in large numbers. Suburbs built in forested areas reduced the numbers of trees and birds, but when plans provided for stands of native vegetation, the diversity of bird species increased as the vegetation grew. In deserts or scrublands, suburban development and artificial watering meant a greater variety of plants and more birds. A study in Arcata, California, by Humboldt State University found that the total number of birds and diversity of the avian population diminished with increased road surface, while the abundance of non-native species increased.

Swainson's Warblers of the southeastern United States have been adjusting to the disappearance of their preferred habitat of canebrakes and swamplands by establishing breeding populations in young pine plantations, which mimic to some degree the thick understory that the birds prefer. For seven to eight years the young pines, destined to become paper pulp, provide a suitable nesting habitat, but once the trees reach 40 feet succession removes the understory and the birds leave. About 60 million acres of pine plantations in various stages of growth are located in the southeastern United States, so some will always be available as nesting sites.

The birds that face the greatest challenges to survival are those with restricted habitats such as the Kirtland's Warbler, which only nests in young Jack Pine stands in a small area of northern Michigan. Once near extinction 50 years ago, the bird population has been brought up to about 5000 birds today thanks to habitat restoration and their range is slowly expanding. The Bali Starling, once almost extirpated, numbered only six birds in 2001. Today about 50 birds live on Bali and another 60 or so on a small Indonesian island nearby where the bird was introduced. Many other threatened bird species occupy small islands. The Floreana Mockingbird, with only 200 individuals, is restricted to two tiny islets in the Galapagos. The Cerulean Paradise-flycatcher exists in a small population of perhaps 100 birds on the island of Sangihe, Indonesia. The islands don't need to be oceanic ones; the Honduran Emerald (hummingbird) lives only in three isolated inland valleys of Honduras.

The take-home lesson is that the outcome of habitat disturbance on avian communities is extremely complex and every situation is unique. We do know, however, that minimizing habitat fragmentation, using environmentally sensitive practices such as the selective logging of trees, leaving edges of native vegetation around farm fields, and planting a variety of shrubbery and trees in suburbs and cities can help considerably in preserving diverse avian communities.

The Bali Starling has started on the road to potential recovery.

CLIMATE CHANGE: HOW BIRDS ADJUST, OR NOT

We know a good deal about the effects of climate change: rising air and ocean temperatures, coral reef bleaching, glacial ice melting, and sea level rising causing the inundation of islands. Lesser-known effects include large swaths of Rocky Mountain forest trees being killed by bark beetles, which now survive the milder winters. It is clear that bird habitats have been affected; the question is if and how birds will survive the effects of increasing temperatures.

A migratory journey initiated by the changing photoperiod has been an appropriate strategy for eons as a lengthening photoperiod in the Northern Hemisphere in the spring paralleled the increasing temperature. Although other factors such as soil moisture come into play, plants begin to leaf out, flower, and produce nectar and fruit as the soil and air warm. Insects and other invertebrates emerge and become active. But rising temperatures and the resultant earlier appearance of flowers and bugs are now causing an increasing disconnect between when the birds arrive and the availability of

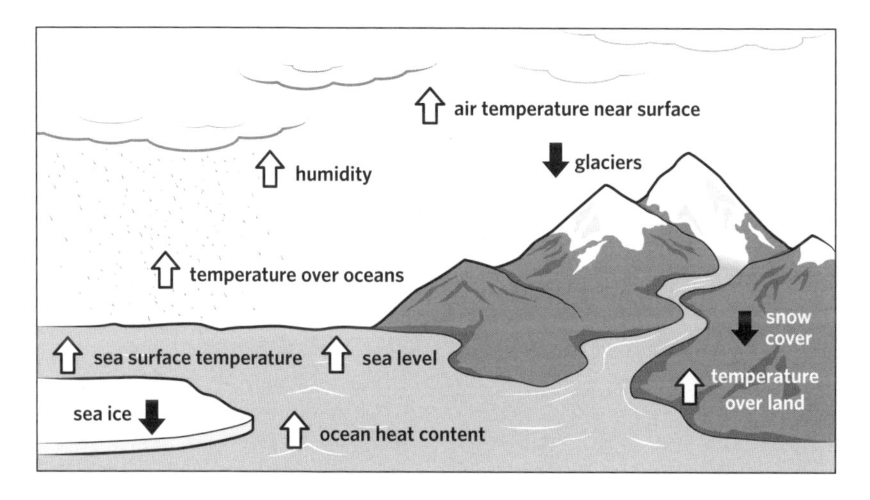

Ten indicators of a warming world.

their food. These changes threaten both immediate and long-term survival of the species as spring is nesting season and abundant food is critical. Birds arriving after the peak of food abundance will face serious hardships, plants that depend upon birds as pollinators get little or no service, and herbivorous insects will have ample opportunities to eat whatever they wish until their aerial predators arrive.

Birds with the longest migration routes or times in passage are at the greatest disadvantage and have suffered greater population declines than resident and short-term migrants. Populations of the European Pied Flycatcher, a black-and-white insectivore that winters in western Africa, have declined up to 90 percent in some locations. Its arrival time at the breeding grounds has not changed and is out of sync with its insect fare, which is emerging earlier. Although migratory behavior is genetically determined, it is malleable and some birds are adjusting. To reach their breeding grounds earlier, individual birds of migratory populations have the flexibility to winter not quite as far south as usual, leave their wintering grounds earlier in the spring, or fly faster to arrive at the nesting area earlier. The Icelandic Black Godwit population has moved the initiation of its northward migration up by two weeks. The birds did not change their spring migration arrival time; instead, they began their breeding cycle earlier than usual because of the warm weather, giving the young a chance to grow and migrate south (to the United Kingdom, Ireland, and France) earlier to find the best wintering areas. These offspring moved northward earlier the next season. So as the older population dies off, succeeding generations will be more in sync with warmer conditions. The American Bird Conservancy says that 20 species of migratory birds in North America arrived three weeks earlier in 2012 than they did in 1965. American Robins in Colorado are now arriving two weeks earlier than they did in 1995. Thirty years worth of data show that birds breeding in Oxfordshire, United Kingdom, have advanced both their arrival and departure times by eight days.

Many birds, including at least seven North American warbler species (Prothonotary, Blue-winged, Golden-winged, Black-throated Gray, Pine, Hooded, and Cape May), have shifted their range northward in the past 24 years by an average of more than 65 miles. The National Audubon Society reports that almost 60 percent of the commoner bird species in North America have relocated their ranges northward by an average of 35 miles and more than 60 species have moved more than 100 miles. The Baltimore Orioles may need a new name because by 2080 their feathered mascot will be nesting in Canada rather than Maryland. Minnesota might require a new state bird as the state is getting too warm for the Common Loon whose entire breeding range is moving to Canada. An exception, the Red-cockaded Woodpecker, has not moved, probably because its inflexible niche restricts it to the pine forests of the far southeastern United States. Thirty-five years of data from the North American Christmas Bird Counts by the Cornell Laboratory of Ornithology indicate that bird species need 30 years or more to shift their ranges in response to warming temperatures.

GLASS COLLISIONS, ALIEN PREDATORS, AND OTHER DILEMMAS

The problems of habitat loss and climate change are growing and seem intractable and perhaps insoluble, at least in the near term. But birds confront plenty of other anthropogenic problems that are significant, but can to some extent be mitigated.

Windows

Have you ever walked into a sliding glass door? Collisions with windows kill at least 100 million birds a year in the United States—more than any other direct cause besides habitat destruction—according to Daniel Klem, the leading expert in bird-window collisions. Not too long ago I received a call from an architectural firm in San Francisco. They were designing a building near the bay with numerous windows, a danger to both resident and migrant birds. Having recognized that problem, the San Francisco Planning Department had earlier established Standards for Bird-Safe Buildings (as New York, Toronto, and Chicago have done). The architects asked me to help them design a building to meet these standards, so I did a lot of research, including talking with Dr. Klem. Ultimately, public opposition to the building quashed the project.

Before 1900, most buildings were mainly brick, stone, wood. With new materials and designs, façades of buildings include a lot more glass. Birds usually cannot see the glass; they either look right through it or they see the reflection of vegetation in it and think there is safe passage. About half of the collisions with windows result in a fatality caused by brain hemorrhage or vertebral damage. We don't know what happens in the long run to those birds that survive the impact and fly off. To reduce bird strikes, one can install screens, put decals or strips of tapes on the glass, hang ornaments in front of the glass, or place a thin film to make the glass opaque to birds. Taking advantage of the fact that birds can see UV, a German firm makes window glass with internal crisscrossing UV strips; birds can see the window but it looks normal to the human eye. Using "fritted" glass with imbedded patterns of lines or dots is another way to make windows more visible to birds. Windows can also be installed at an angle with the bottom of the glass further back than the top. This simultaneously minimizes the force of the impact as the bird doesn't hit the glass straight on and the ledge formed at the bottom of the window provides a place for a stunned bird to recover.

Bird-window strikes can happen anywhere at any time, but most collisions are with large windows near ground level. Migration time is dangerous because birds are on the move, but so is wintertime for those birds that come to feeders. If you have a feeder, either place it within at least 3 feet of the

University of Toronto buildings with fritted glass to reduce bird strikes.

house so the birds can't pick up speed when they leave, or far away from the house so it's not near any windows.

According to Dr. Klem, a number of rare birds are further imperiled by window collisions: the Swift Parrot of Australia, the Cerulean Warbler of North America (under consideration as an endangered species), the Black Rail of the southeastern United States and the Caribbean, the Kirtland's Warbler of Michigan, and the Plain Pigeon of the West Indies. Unlike habitat destruction and climate change, individual and collective solutions exist for buildings and their windows. Birds have no survival skills to avoid windows without some help from us so implementing bird-safe solutions is imperative.

Cats and Other Alien Predators

The Egyptians domesticated house cats about 4000 years ago, revered them as sacred, and mummified them when they died (one cat cemetery in ancient Egypt contained 300,000 cats). Cats reached the Middle East and Europe about 1000 BC on the ships of Greek and Phoenician traders as the felines were kept on ships to control vermin. Roman armies carried cats with them to protect their grain supplies and when the armies retreated the cats remained in the United Kingdom and parts of Europe. House cats became common in the United States in the 19th century to keep agriculture-loving mice and rats under control.

An estimated 77 million domestic cats live in the United States today (maybe 600 million worldwide) along with an unknown number of feral felines. According to a U.K. study, the density of cats may be up to 665 per square mile, this number negatively correlating with the number of bird species. Only one-third of cat owners keep their cats exclusively indoors, so they hunt outdoors. In a U.S. study, small cameras attached to house cats that were free to roam indicated that they ate perhaps a billion birds each year. Add in feral cats and the number rises to 4 billion. Figures for the United Kingdom, Australia, and New Zealand are equally dismal. Perhaps 33 bird species worldwide have been exterminated by cats. The only positive side to this tale is that felines tend to take a decent proportion of less fit birds, many of which would not survive anyway.

This problem doesn't have any easy solutions, but disposing of feral cats is a good start. My local city park was an unwanted pet dumping ground for years. On my walks I often spotted dozens of feral cats, especially adolescents, and noted that certain birds such as the California Quail had disappeared from the park, and birds of the lower vegetation such as towhees, thrushes, and robins were declining precipitously. At the same time cat aficionados were putting food out for the stray felines, no doubt increasing their numbers. Happily the volunteer organization Cat Coalition began trapping cats and finding homes for them. More than 1000 cats were removed from the park and after a few years quail and other birds reappeared.

Unfortunately, feral cats still roam my local park, as they do in many areas worldwide. There's a story about the Stephen's Island Wren, a flightless songbird endemic to the island of the same name near New Zealand. The birds were rare and the lighthouse keeper's cat contributed to the bird's extinction around 1900. About 1870, another New Zealand island, Little Barrier, was invaded by felines that went on to play a significant role in the total extinction of the Little Barrier Snipe, the local extinction of the North

Little Barrier Snipe, painted in 1844, would be extinct by about 1870.

Island Saddleback, and the severe reduction in numbers of the Grey-faced, Cook's, and Black Petrels. In 1905 New Zealand's *Canterbury Press* had this quote: "And we certainly think that it would be as well if the Marine Department, in sending lighthouse keepers to isolated islands where interesting specimens of native birds are known or believed to exist, were to see that they are not allowed to take any cats with them, even if mouse-traps have to be furnished at the cost of the state."

Cats are not the only alien invader to threaten birds. Dogs, goats, rats, snakes, opossums, and other predators have been major causes of extinction on oceanic islands. Virtually every native bird species such as the Cardinal Honeyeater and Guam Flycatcher was extirpated from the island of Guam by the brown tree snake, accidentally introduced with shipments of lumber from New Guinea just after World War II. Finding the island teeming with birds and lizards, the snakes advanced at a mile a year.

A study of 220 islands revealed that each successive predator introduction increased the number of endemic bird species lost. As endemic species declined and alien birds became more common, extinction rates fell. The more adaptable alien birds were apparently able to withstand pressure from predators. This is a lesson for those islands which still have endemic species not yet threatened by predators—do not let any more in.

Additional Hazards

Humans have created a number of other hazards that birds face on a regular basis all over the world. An unfortunate event occurred on New Year's Eve in 2010 in Beebe, Arkansas. Thousands of Red-winged Blackbirds were roosting in a wooded area on the edge of town when, about 11:30 P.M., they came flying chaotically into town, frightened and disoriented, colliding with buildings, signs, cars, and each other; 2000–3000 birds died. Speculation as to the cause of death ranged from poisoning to a UFO invasion, but the most likely explanation is that they died of blunt force trauma or hypothermia or both. Huddled together for warmth for the night, the diurnal birds were driven by the noise of fireworks into the cold darkness and their demise. Similar mass bird deaths occurred in the week following that event; Turtle Doves died in Italy, Jackdaws in Sweden, geese and ducks in Ontario, and more Red-winged Blackbirds in Louisiana. It appears that all the birds were frightened by noisy celebratory activity and driven out of their roosts to their deaths.

I could write many more pages about problems such as light, noise, pollution, pesticides, power lines, communication towers, wind turbines, automobiles, airplanes, oil spills, fishing by-catch, hunting, poaching, egg-collecting, the pet trade, off-road vehicles, imported diseases, and lead ingestion, each of which contribute to the demise of birds. Enumerating bird deaths due to various causes is difficult and the numbers vary widely depending on the methodology used to collect data. Solar panels, for example, are estimated to kill 1000–28,000 birds a year and wind turbines 100,000–300,000. Coal may be responsible for eight million avian fatalities a year if the production and transportation of coal as well as its contribution to global warming are figured in. But, again, the two overwhelming causes for the reduction in bird populations are habitat destruction and climate change, followed by collisions with buildings and power lines, and predation by cats. We could dwell further on these awful tribulations, but let's instead examine some of the ways that birds are adapting to the changing environment.

BIRDS ADAPTING TO A NEW WORLD

Urbanization—a shift from rural to urban areas—is perhaps the most difficult challenge to bird survival and may be a major driver of avian species extinction in this century. Only 2.7 percent of the earth is covered by cities yet the majority of the human population lives in them. The United Nations predicts that by 2050 more than half the population of the world will live in cities. In addition to habitat disruption and disappearance, the effect of urbanization is intense because of all it encompasses. Cities mean people, and the more people and their trappings, the fewer birds. During a weeklong visit to New York City, I saw few birds and virtually all of them were pigeons or House Sparrows. The two exceptions were in Central Park: a very cold Blue Jay on a fence and a soaring Red-tailed Hawk. Cities are a drastic change from native ecosystems.

Realizing that urbanization is spreading rapidly and changing habitats for birds, some ornithologists are shifting their research focus as well. Since 1970, the number of published research papers on urban birds and their habitats has more than quadrupled. The vast majority of studies have been conducted in North America and Europe; fewer than 5 percent were done in developing countries. Given that the population and thus urbanization of developing countries is happening at a far faster rate than in developed countries, the survival of birds faces major threats without the information needed to protect them.

But birds are not disappearing under asphalt and bricks without a fight. Many birds have adjusted to, and even benefit from, human activity. A good number have adapted to survive city life, particularly Rock Doves, House Sparrows, and European Starlings. In fact, more than 20 percent of the world's bird species are represented in cities. Caracas, Venezuela, a city of six million people, has a non-native population of Blue and Gold Macaws surviving on what they can forage and what the locals leave out for them. Local legend has it that an Italian immigrant, Vittorio Poggi, rode around the city on his motorcycle with a macaw he trained to fly alongside; he eventually released 100 more birds which he bought, bred, or received from people who no longer wanted them as pets.

South London hosts several colonies of wild parrots, the most common being Rose-ringed Parakeets with perhaps 8200 pairs. Native to parts of Asia and Africa, Rose-ringed Parakeets were once kept by ancient Greeks and Romans and are popular as pets today. Their population has been on the rise at least since the 1990s, having escaped either from a broken shipping crate at Heathrow Airport or from the collapse of a large aviary, or having been released by the late musician Jimi Hendrix on Carnaby Street. The long, chilly winters no doubt cull the bird's populations but they are surprisingly hardy, feeding on berries, seeds, nuts, buds, vegetables, and fruit. A bit aggressive, they tend to dominate bird feeders. Everyone seems to have adjusted to them and the U.K. Department of Environment, Food, and Rural Affairs (DEFRA) has deemed their elimination not cost effective. The story of the Monk Parakeet is different. Native to Brazil and Bolivia, they became established in London, probably because of escapees or releases from the pet trade. Their enormous communal stick nests, some as large as a small car, are often built on power transmission poles or towers and when wet sometimes

The Macaws of Caracas, Venezuela.

cause blackouts or fires. Although the Monk Parakeet population is only about 150 birds, they pose such hazards that DEFRA has decided to reduce the population to 50 birds.

Parrots are common in other cities as well. In the center of San Jose, Costa Rica, amid buildings, shoppers, traffic, and general bustle, hundreds of Red-lored Parrots adorn building ledges and the trees of the central park. San Francisco is known for its flock of Red-masked Parakeets on Telegraph Hill, a bird colony that was generated, apparently, by a couple of escapees. The city of Bakersfield, California, hosts 1000 Rose-ringed Parakeets. Large flocks of Red-lored Parrots, Mitred Parakeets, and Lilac-crowned, Blue-fronted, White-fronted, and Yellow-headed Amazons can be found in urban locales of southern California where they live off the fruits and nuts of plants, exotic imports like the birds themselves. Non-native parrots of three dozen species have found refuge in North America, including the very adaptable Monk Parakeets of Brooklyn, Chicago, and Montreal.

Overall, urban bird populations average 30 percent higher than nearby rural populations of the same species. Feral pigeons and House Sparrows have been associated with humans for thousands of years so it's not too

A potential threat to crops, the Monk Parakeet population is increasing rapidly in the United States and its large nests have interfered with electrical transmission lines.

surprising that the populations of these birds are higher in urban habitats as compared to nearby rural areas. But even European Blackbirds, which invaded cities less than two hundred years ago, have urban populations two orders of magnitude higher than adjacent woodlands, their native habitat. Since the 1950s, 47 bird species living in the inner city of Warsaw, Poland, have diminished in number or disappeared as the city grew, but the populations of 37 species increased and 12 new bird species colonized the city.

About 114 bird species utilize cities or suburbs worldwide for part or all of their lifespan. Although domestic predators (dogs and cats) are common, predation pressure is low because dogs and cats also have human food to rely on and birds have an infinite number of hiding places. Gulls, vultures, ravens, and crows hang out at garbage dumps and along roadways looking for the superabundant food scraps. Hummingbirds, towhees, chickadees, nuthatches, sparrows, doves, finches, jays, and woodpeckers frequent bird feeders. Falcons, hawks, and owls even nest in cities, where they can find abundant prey. George and Gracie, a Peregrine Falcon pair, have been nesting on a San Francisco building since 2005; in 2014 they nested in a 30th-floor planter on a financial district skyscraper. A famous Red-tailed Hawk in New York City, dubbed Pale Male because of his light-colored head, has nested on an apartment building overhang on Fifth Avenue for 24 years with a new female about every three years. That he has survived that long is amazing as 75 percent of Red-tailed Hawks die before their first birthday.

The populations of year-round city resident birds have held steady or showed a strong increase since the mid-1970s. However, migratory birds that only use urban areas for nesting, such as Common Nighthawks and Chimney Swifts, are declining because competition for food and nesting sites with the growing populations of sedentary urban birds is too great for them. As cities grow, fewer and fewer niches remain for birds that only visit for a few months of the year.

New Reactions to City Predators

Another indication that birds are adapting to urban environments is their reaction to city-living predators. An analysis of the response of 44 European bird species to an approaching human showed that urban birds flew shorter distances to escape threats than rural birds of the same species. The longer the bird species were associated with the city (as judged by the number of generations) and the larger their populations, the shorter the escape distances, indicating the birds are adapting to city life. Urban birds with shorter flight distances are also less susceptible to predation by the European Sparrowhawk (and presumably other predators). However, as the prey population increased, so did the success of the sparrowhawks as they honed their urban hunting skills.

Besides cats, mammalian predators in urban and suburban areas include raccoons, opossums, foxes, dogs, rats, and skunks. Birds have few survival tactics except mobbing or escape to cope with predators, but researchers in Europe say that might be changing. They captured 1132 birds of 15 species in rural and urban areas, grading every bird on its escape behavior when in a human hand—squirming, biting, alarm calls, and feather loss. Researchers also timed how long each bird remained immobile "playing possum"—when laid flat in a human's open palm. Urban birds shed more feathers, let out more calls and screams, and flew from the hand faster than country birds did. It appears that these tactics are escape techniques modified to cope with city predators, especially cats.

Plumage Color Changes

Originally from Asia, the highly invasive Morrow's Honeysuckle has become common in the suburbs of the northeastern United States. Cedar Waxwings are frugivorous birds with yellow-tipped tail feathers. After about 1950, birds were being found with orange-colored feather tips, a result of the birds ingesting honeysuckle berries with the red-purple pigment rhodoxanthin. Whether this color change has any social or sexual implications for the birds is not known. In central Ohio, the brightness of male Northern Cardinals' red plumage decreased as urbanization and exotic honeysuckle plants increased. The brighter red males from rural or forested areas with little or no honeysuckle bred earlier and raised more offspring than their dull city dweller cousins.

In Europe, the favorite food of the Great Tit is caterpillars, which live on a diet of carotenoid-containing leaves. Carotenoids are important as precursor vitamins, antioxidants, and pigments in birds, responsible for the Great Tit's yellow chest and abdomen. Urban-dwelling caterpillars are less abundant, find fewer leaves, and contain lower amounts of carotenoids, so parent Great Tits have to make more foraging trips to feed their young caterpillars than forest-inhabiting nesters. In a polluted area, the carotenoid levels of the insects are so low that the young birds leave the nest with a dull yellow chest rather than a bright one. Whether these changes in feather color will alter social interactions among the birds is yet to be seen.

Researchers in Paris discovered that darker city (feral) pigeons are healthier than lighter ones, partly because darker birds can rid their bodies of heavy metals. Wild birds kept in cages for a year had their flight feathers tested for heavy metals such as zinc, cadmium, copper, and lead. Eumelanin, the pigment responsible for the black-brown dark-colored feathers, binds to heavy metals. At the next molt the new feathers were tested for heavy metals; the levels were down by 75 percent. Some evidence shows that darker pigeons are increasing as a proportion of the population, because their survival rates are higher and reproduction more successful. Like many evolutionary changes the darker birds just happened to have a physiology that benefited them when they inhabit an urban area with heavy metals.

Nesting and Foraging Modifications

The Barn Swallow has been associated with human environments in Europe for more than 4000 years and today over 99 percent of Barn Swallows nest indoors in barns, sheds, or other buildings. Given a generation time (average time between two consecutive generations) of Barn Swallows of 1.59 years, about 2500 generations allowed plenty of time to adapt to nesting in warmer, safer, human-constructed places. Chimney Swifts are so accustomed to nesting in human-made structures that only 10 instances of the bird nesting in the wild in the last century are known. The European Blackbird has only been nesting in close proximity to humans for about a hundred years. With a generation time of 2.27 years, they have only had 44 generations to adapt, so only 15 percent of blackbirds nest indoors. But city-dwelling blackbirds nest a few weeks earlier than their rural relatives and have two or three broods compared to the single brood of those nesting in forests; there have even been instances of winter broods among city blackbirds. The increased amount of light, providing more time for foraging and inducing the birds to breed earlier in the season, amelioration of weather conditions, and both natural and human food sources apparently make life a bit easier for these urban blackbirds.

Some wild bird species incorporate aromatic plants into their nests to repel parasites. House Sparrows and House Finches in Mexico City put fibers from filter cigarette butts into their nests. Nicotine is a known arthropod repellent and the cellulose acetate in the filters contains significant amounts of the chemical. The more cellulose acetate in the nest the fewer the mites infest it. The cellulose acetate may simply be available nesting material for the birds, but the birds are benefitting from its parasite-repellent qualities. Although this study was done in 2013, there have been reports of birds incorporating cigarette filters into their nests for at least 45 years. One downside of this behavior is reflected in the incident of a building fire in South London, caused by a bird carrying a lit cigarette butt into its nest in a hole in the roof of the building. Several other reports from fire departments indicate that birds such as House Sparrows, European Starlings, and Rock Doves have been implicated in starting structural fires.

House Finch nest.

Birds learn the characteristics of their city with its litter, garbage, and intentional feeding providing ample opportunity for foraging. House Sparrows and starlings frequent outdoor cafes, pigeons extend their feeding bouts by foraging under streetlights at night, and many species quickly learn the location of bird feeders. In 1994 the month of February was designated as the official bird-feeding month in the United States. Over 60 million Americans feed birds at various kinds of feeders, spending over four billion dollars annually for food and feeders. Dozens of different species and billions of birds partake in these backyard offers. Bird feeding has been a regular pastime for at least a century and a half and probably far longer.

Adapting to Traffic

A 2013 paper reported on a 30-year study of Cliff Swallows killed by vehicles in Nebraska. Colonies nesting on bridges and overpasses flew over highways and passing traffic killed the birds. In the 1980s the kills averaged 20 birds per year, but since 2010 the annual average had dropped to five birds even though the colony had doubled in size and vehicular traffic stayed steady.

Evening Grosbeak at feeder.

Researchers Charles and Mary Brown wondered what was happening, so they took measurements of the deceased birds and compared them to live birds in the colony. The live birds had a 4.2-inch-long wing while the wings of the road-killed birds were 4.5 inches long. Studies of swallow flights indicate that shorter wings are better for making quick turns and make the birds better at dodging traffic; "natural" selection (by vehicles) is weeding out the longer-winged birds.

Researchers in France measured the distance from moving vehicles at which birds began flying to avoid being hit. One would think that the faster the vehicle is traveling, the greater the distance at which the bird would initiate flight. Actually, the birds responded not to the speed of the vehicle, but to the posted speed limit! It appears that birds that regularly inhabit a stretch of roadway have learned the most common speed, which is somewhere around the posted limit, and react as if the vehicle was traveling at that pace, even if it were dawdling or speeding. Since most cars travel around the posted limit, this appears to be the best strategy to avoid collisions with oncoming traffic.

Noise and Bright Lights

Songbirds convey information to mates and competitors through song, but as environments get noisier, communication becomes more difficult. Long-term survival dictates some adjustment. Because most city noise (traffic, lawn mowers, industry) happens at low frequencies and the bird songs overlap some of those frequencies, it is advantageous for the singers to emphasize higher frequency song notes and dampen their low frequency notes. Male Song Sparrows in Portland, Oregon, sing higher-frequency notes and reduce the loudness of their low notes in noisy city environments. European Blackbirds in Salamanca, Spain, act similarly—more ambient noise causes birds to increase the pitch of their song. In the city with a noise level of 66 decibels, the average maximum frequency of bird songs was 3165 Hz while in the rural surrounding rural area where noise levels averaged only 37 decibels the song frequency was 2657 Hz. Comparable effects have been observed in Great Tits and Common Nightingales. Is this an evolutionary adjustment or just an acclimation by individual birds? Two populations of the Reed Warbler, widespread across Europe, were studied. One population resided in an area that was noisy for much of the year; the other lived in a much quieter locale. In the noisy environment, the Reed Warblers sang higher frequency songs than the birds in the quiet location. But when the noise levels were similar, so were the song frequencies. This indicates flexibility in their song repertoire, not necessarily an evolutionary change. But in most songbirds, a significant learning component by young birds occurs about a year after fledging. So it could very well be that young birds in noisy environments will only learn the higher ranges of their species' songs and pass this knowledge onto their offspring. Being able to communicate one's song is necessary to survival, so shifts in genetic makeup are likely to occur. Birds like the Rufous-crowned Sparrow simply decline or disappear with urbanization, so their songs remain unaffected in wild populations.

Since the arrival of the first commercially successful light bulb in 1879, the night sky has increased in brightness by a factor of six thousand. Scientists in Vienna found that in areas with streetlights, some species of birds begin to sing earlier than they do under normal lighting. In a seven-year study of the reproductive behavior of Blue Tits in the United Kingdom, female birds in

Rufous-crowned Sparrow singing.

nests on the forest edge near streetlights began to lay eggs an average of 1.5 days earlier than those in darker areas. Males in lighted areas were twice as successful in mating with a female in addition to his mate. It seems that the additional light gave the females more time to prepare the nest and start laying eggs and the males had additional time to sing and attract females from a different territory. Studies in Leipzig, Germany, found that the European Blackbird is healthier in cities than in the wild because the near-constant light of the city allows it to forage a lot longer.

Migration

Adapting to urban environments apparently means adopting a sedentary lifestyle. The European Blackbird was formerly a migratory bird that preferred forests, but over the last several decades it has become a common city bird. Researchers compared fourteen populations of blackbirds, seven urban and seven rural, across a broad swath of northern Europe and found a strong trend for the urban populations to be more sedentary than rural ones. The tendency was strongest in the more northerly populations of Latvia and Estonia, even though the colonization by blackbirds in those countries was rather recent, occurring somewhere between 1930 and 1950, indicating a fairly rapid evolution of sedentary behavior. But a number of migratory bird species continue to nest in cities, such as the Pied Wagtail, Willow Warbler, and Song Thrush of Europe, and the American Robin and Cliff Swallow of North America.

Various urban features, too many to expound on here, affect migratory birds, but a particularly interesting one is all the electronic emanation from cell phones, radios, TV towers, and satellites—a common feature of cities. We don't know much about how birds are affected by all this but one study by a German researcher, Henrik Mouritsen, who moved his laboratory from a rural area to the city of Oldenburg, gives us some indication. After moving, he found that the European Robins he was investigating no longer oriented in the proper direction in their wooden cages. After three years he discovered the cause was electronic interference from the outside. It wasn't mobile phones, power lines, or wireless signals, but AM frequencies, those used by radio stations and certain kinds of electronic equipment. When he used aluminum-screened cages and grounded them electrically, the birds oriented normally. AM frequencies are stronger and reach farther at night, which may affect the birds' navigational abilities. The study shows us that there are all kinds of electronic signals in our urban world and we have much to learn about how they affect the lives of birds and whether they are adapting to them in some way.

PEOPLE HELPING BIRDS SURVIVE

Birds cannot survive in the long run without assistance from us in the form of various organizations dedicated to birds. In the mid-18th century, Marie Antoinette was a fashion trendsetter, which eventually led to her downfall from the French throne. By wearing feathers in her hat she initiated a demand for feathers that by the 1850s was in full swing with hundreds of thousands of birds being killed for their plumes, especially egrets. In the United Kingdom at that time, the price of feathers by weight was equal to that of gold. Songbirds and even entire Arctic Terns were mounted on hats and sizeable colonies of birds were decimated in the process. Just as egret colonies teetered on the brink of extinction, the Lacey Act, prohibiting the interstate commerce of wildlife, was enacted in the United States in 1900, ending the bulk of the commercial plume trade.

The National Audubon Society of the United States came into being to stop the slaughter of wild birds and today is involved in many major protection efforts across the country. The Royal Society for the Protection of Birds was formed in the United Kingdom to counter the feather trade in the Victorian era and today focuses on conserving and protecting natural habitats. BirdLife International is the world's largest nature conservation coalition, with 120 partners across the world who manage 1553 reserves or protected areas covering 11.1 million acres.

Numerous programs are aimed at teaching people to appreciate birds, value the natural world, and support efforts to protect and preserve its inhabitants. The Celebration of Urban Birds, the Urban Bird Sounds Project, Neighborhood Nestwatch, the Fledgling Birder's Institute, the Birding Challenge, National Wildlife Federation Schoolyard (and Backyard) Habitats, and the Shorebird Sister Schools Program are just a few. The International Migratory Bird Day highlights and celebrates the migration of nearly 350 species of migratory birds between nesting habitats in North America and their wintering grounds. The Bird Education Alliance for Conservation is a coalition of educators representing universities, bird observatories, local, state, and federal agencies, and environmental education and conservation groups. Partners in Flight is a cooperative among federal, state, and local governments, charities, professional organization, conservation groups, industry, and others to seek ways to protect migratory land birds. And the list happily continues.

Birdwatchers: The Ultimate Key to Survival

Birdwatching is the fastest growing outdoor activity; somewhere between 45 and 85 million people call themselves birdwatchers and invest \$20–\$35 billion each year in birding activities. The average birdwatcher has an above average income and spends \$1500–\$2000 a year on books, binoculars,

The Blue Jay is a bird most people are familiar with and no doubt is responsible for getting people interested in nature and conservation.

telescopes, travel, and lodging. Serious birders might spend close to $5000 for an exotic birding trip led by one of 127 bird-guiding companies. Neil Hayward, from Cambridge, United Kingdom, sighted 750 species in North America in 2013. He spent 195 nights away from home, drove 51,758 miles, was at sea for 147 hours, and took 177 flights through 56 airports while on his quest.

More and more bird festivals are offered each year, and big days and big years are organized all the time. January 5 is National Bird Day. It's a big sport, but not just an esoteric one; birdwatching is becoming more ingrained into our national consciousness. But most birdwatchers don't participate in festivals or competitions, they just want to incorporate birdwatching into their everyday strolls.

Too often there is a disconnect between the natural world of birds and other organisms and our technology-obsessed everyday life. Birdwatching connects people to the environment and helps them understand why birds need protection from human activities. And the more we understand birds, the more we understand the environment from which they come

Birdwatchers play a key role in helping birds survive.

and which they depend upon for survival, and the more likely we are to be able to protect them.

Until the early 20th century, birds were being slaughtered for food, feathers, eggs, and sport. The Passenger Pigeon, Labrador Duck, and Great Auk did not survive this onslaught, but herons, egrets, and songbirds did thanks to the efforts of some enlightened politicians and activist citizen groups. From the role of protecting birds, the mission of protecting ecosystems arose slowly but necessarily. Are these efforts and laws sufficient to ensure the survival of birds? I am cautiously optimistic. The individuals, agencies, and industries that have created the greatest challenges to the survival of birds—habitat destruction and climate change—tend to be shortsighted and have goals at odds with bird conservation. On the other hand, the powerful plumage trade, the egg and meat hunters, and overhunting in the name of sport have largely been halted, so change is possible. Political action is necessary, but education is primary. Birds have evolved superb adaptations to the natural world and evolution polished those adaptations as the world changed. But now that we humans have put environmental change in high gear, birds can no longer keep up, so we need to be partners in their survival.

METRIC CONVERSIONS

INCHES	CM
1	2.5
2	5.1
3	7.6
4	10
5	13
6	15
7	18
8	20
9	23
10	25
11	28
12	30
13	33
14	36
15	38

FEET	M
1	0.3
2	0.6
3	0.9
4	1.2
5	1.5
6	1.8
7	2.1
8	2.4
9	2.7
10	3
20	6
40	12
60	18
80	24
100	30
130	40
600	182
5200	1600
11,200	3400
30,000	9144

MILES	KM
1	1.6
2	3.2
3	4.8
4	6.4
5	8.0
6	9.7
7	11
8	13
9	14
10	16
15	25
20	32
30	48
40	66
50	80
60	97
70	113
80	129
90	145
100	161
200	322
300	483
400	644
500	805
600	966
700	1127
800	1287
900	1448
1000	1609
3000	4828
6000	9656
50,000	80,467

OUNCES	GM
0.05	1.4
0.175	5
0.5	14
6	170

POUNDS	TONS	KG
0.5		0.23
1		0.45
1.5		0.68
2		0.91
4		2
20		9
40		18
100		45
150		68
200		91
300		140
400		180
500		230
600		270
700		320
800		360
900		400
1000		460
2000	1	910

TEMPERATURES
°C = 5/9 × (°F – 32)
°F = (9/5 × °C) + 32

CELSIUS	FAHRENHEIT
-10	14
-9	15.8
-8	17.6
-7	19.4
-6	21.2
-5	23
-4	24.8
-3	26.6
-2	28.4
-1	30.2
0	32
1	33.8
2	35.6
3	37.4
4	39.2
5	41
6	42.8
7	44.6
8	46.4
9	48.2
10	50
20	68
30	86
40	104
50	122
60	140

BIBLIOGRAPHY

BIRDS, BEAKS, AND BELLIES

Aristotle. 2001. *On the Parts of Animals.* Trans. J. G. Lennox. Oxford: Clarendon Press.

Baird, J. 1980. The Selection and Use of Fruit by Birds in an Eastern Forest. *The Wilson Bulletin* 92 (1): 63–73.

Beecher, N. A., R. J. Johnson, J. R. Brandle, R. M. Case, and L. J. Young. 2002. Agroecology of Birds in Organic and Nonorganic Farmland. *Conservation Biology* 16 (6): 1620–1631.

Benkman, C. W. 1988. Seed Handling Ability, Bill Structure, and the Cost of Specialization for Crossbills. *The Auk* 105 (4): 715–719.

Birkhead, T. R., J. Wimpenny, and R. D. Montgomerie. 2014. *Ten Thousand Birds: Ornithology since Darwin.* Princeton, NJ: Princeton University Press. An excellent view of the history of ornithology.

Bock, W. J. 1966. An Approach to the Functional Analysis of Bill Shape. *The Auk* 83 (1): 10–51. A classic study.

Bull, E. L., and R. C. Beckwith. 1993. Diet and Foraging Behavior of Vaux's Swifts in Northeastern Oregon. *The Condor* 95 (4): 1016–1023.

Caceci, T. 2013. Avian Digestive System, 2013. In VetMed Veterinary Histology Web. vetmed.vt.edu/education/curriculum/vm8054/Labs/Lab22/lab22.htm.

Crome, F. 1985. An Experimental Investigation of Filter-Feeding on Zooplankton by Some Specialized Waterfowl. *Australian Journal of Zoology* 33 (6): 849–862.

Cunningham, S., I. Castro, and M. Alley. 2007. A New Prey-detection Mechanism for Kiwi (*Apteryx* spp.) Suggests Convergent Evolution between Paleognathous and Neognathous Birds. *Journal of Anatomy* 211: 493–502.

Diaz, M. 1994. Variability in Seed Size Selection by Granivorous Passerines: Effects of Bird Size, Bird Size Variability, and Ecological Plasticity. *Oecologia* 99 (1–2): 1–6.

Díaz, M. 1996. Food Choice by Seed-eating Birds in Relation to Seed Chemistry. *Comparative Biochemistry and Physiology Part A: Physiology* 113 (3): 239–246.

Dickson, J. G. 1979. Stephen F. Austin State University, School of Forestry, Proceedings of the symposium, The Role of Insectivorous Birds in Forest Ecosystems. New York: Academic Press.

Ehrlich, P. R., D. S. Dobkin, and D. Wheye. 1988. *The Birder's Handbook: A Field Guide to the Natural History of North American Birds: Including All Species That Regularly Breed North of Mexico*. New York: Simon & Schuster. A nice overview of the life histories of North American birds.

Fenton, M. B., and T. H. Fleming. 1976. Ecological Interactions between Bats and Nocturnal Birds. *Biotropica* 8 (2): 104–110.

Galetti, M., R. Guevara, M. C. Cortes, R. Fadini, S. Von Matter, A. B. Leite, F. Labecca, T. Ribeiro, C. S. Carvalho, R. G. Collevatti, M. M. Pires, P. R. Guimaraes, P. H. Brancalion, M. C. Ribeiro, and P. Jordano. 2013. Functional Extinction of Birds Drives Rapid Evolutionary Changes in Seed Size. *Science* 340 (6136): 1086–1090.

Gill, F. B. 1990. *Ornithology*. New York: W.H. Freeman.

Gill, F. B., and D. Donsker, eds. 2015. IOC World Bird List (v 5.2). worldbirdnames.org/.

Handbook of the Birds of the World. Vols 1–16. 1992–2011. Barcelona: Lynx Edicions.

Hanson, T. 2011. *Feathers*. Philadelphia: Basic Books.

Herd, R., and T. Dawson. 1984. Fiber Digestion in the Emu, *Dromaius novaehollandiae*, a Large Bird with a Simple Gut and High Rates of Passage. *Physiological Zoology* 57 (1): 70–84.

Herrera, C. M. 1984. A Study of Avian Frugivores, Bird-Dispersed Plants, and Their Interaction in Mediterranean Scrublands. *Ecological Monographs* 54 (1): 1–23.

Kricher, J. C. 1997. *A Neotropical Companion: An Introduction to the Animals, Plants, and Ecosystems of the New World Tropics*. Princeton, NJ: Princeton University Press.

Lederer, R. J. 1972. The Role of Avian Rictal Bristles. *The Wilson Bulletin* 84: 193–197.

Lederer, R. J. 1975. Bill Size, Food Size, and Jaw Forces of Insectivorous Birds. *The Auk* 92 (2): 385–387.

Levey, D. J. 1987. Seed Size and Fruit-Handling Techniques of Avian Frugivores. *The American Naturalist* 129 (4): 471–487.

Markula, A., M. Hannan, and S. Csurhes. 2009. Invasive Plants and Animals—The Red Quelea. Queensland Primary Industries and Fisheries, *Biosecurity Queensland*: 1–13.

Mason, J. R., and L. Clark. 2000. The Chemical Senses in Birds. In *Sturkie's Avian Physiology*, 5th ed. Edited by G. C. Whittow. New York: Academic Press. 39–56.

Nebel, S., D. L. Jackson, and R. W. Elner. 2005. Functional Association of Bill Morphology and Foraging Behaviour in Calidrid Sandpipers. *Animal Biology* 55 (3): 235–243.

Peters, V. E., R. Mordecai, C. R. Carroll, R. J. Cooper, and R. Greenberg. 2010. Bird Community Response to Fruit Energy. *Journal of Animal Ecology* 79 (4): 824–835.

Proctor, V. W. 1968. Long-Distance Dispersal of Seeds by Retention in Digestive Tract of Birds. *Science* 160 (3825): 321–322.

Pyke, G. H., H. R. Pulliam, and E. L. Charnov. 1977. Optimal Foraging: A Selective Review of Theory and Tests. *The Quarterly Review of Biology* 52 (2): 137–154. A well-known review paper on ideas of foraging.

Shrestha, M., A. G. Dyer, S. Boyd-Gerny, R. B. M. Wong, and M. Burd. 2013. Shades of Red: Bird-pollinated Flowers Target the Specific Colour Discrimination Abilities of Avian Vision. *New Phytologist* 198 (1): 301–310.

Sibley, D. 2000. *The Sibley Guide to Birds*. New York: Alfred A. Knopf.

Steiger, S. S, A. E. Fidler, M. Valcu, and B. Kempenaers. 2008. Avian Olfactory Receptor Gene Repertoires: Evidence for a Well-developed Sense of Smell in Birds? *Proceedings of the Royal Society B: Biological Sciences* 275 (1649): 2309–2317.

Svensson, L., P. J. Grant, K. Mullarney, D. Zetterstrom, and D. A. Christie. 1999. *The Complete Guide to the Birds of Europe*. Princeton, NJ: Princeton University Press.

The Peregrine Fund. peregrinefund.org.

The Raptor Trust. theraptortrust.org.

Tucker, V. A. The Deep Fovea, Sideways Vision and Spiral Flight Paths in Raptors. 2005. *Animal Biology* 55 (3): 235–243. Tucker is a pioneer in avian flight research.

Valdés-Peña, R. A., S. Gabriela Ortiz-Maciel, S. O. Valdez Juarez, E. C. E. Hoeflich, and N. F. R. Snyder. 2008. Use of Clay Licks by Maroon-Fronted Parrots (*Rhynchopsitta terrisi*) in Northern Mexico. *The Wilson Journal of Ornithology* 120 (1): 176–180.

Van der Meij, M. A. A., and R. G. Bout. 2006. Seed Husking Performance and Maximal Bite Force in Finches. *The Journal of Experimental Biology* 209: 3329–3335.

Vander Wall, S. 1990. *Food Hoarding in Animals*. Chicago: University of Chicago Press.

Veterinary Histology. 2014. *VM8054: Veterinary Histology*. Virginia-Maryland Regional College of Veterinary Medicine.

Waldvogel, J. 1990. A Bird's Eye View. *American Scientist* 78 (4): 342–353.

Wheelwright, N. T., and C. H. Janson. 1985. Colors of Fruit Displays of Bird-Dispersed Plants in Two Tropical Forests. *The American Naturalist* 126 (6): 777.

Withers, W. 1983. Energy, Water, and Solute Balance of the Ostrich *Struthio camelus*. *Physiological Zoology* 56 (4): 568–579.

Zweers, G., F. De Jong, and H. Berkhoudt. 1995. Filter Feeding in Flamingos (*Phoenicopterus ruber*). *The Condor* 97 (2): 297–324.

CAN YOU SEE UV?

Altshuler, D. L. 2001. Observational Learning in Hummingbirds. *The Auk* 118 (3): 795–799. Study reports on hummingbirds learning from others at hummingbird feeders.

Benites, P., M. D. Eaton, D. A. Lijtmaer, S. C. Lougheed, and P. L. Tubaro. 2010. Analysis from Avian Visual Perspective Reveals Plumage Colour Differences among

Females of Capuchino Seedeaters (*Sporophila*). *Journal of Avian Biology* 41 (6): 597–602.

Birkhead, T. R. 2012. *Bird Sense: What It's like to Be a Bird*. New York: Walker Books.

Bize, P., R. Piault, B. Moureau, and P. Heeb. 2006. A UV Signal of Offspring Condition Mediates Context-dependent Parental Favouritism. *Proceedings of the Royal Society B: Biological Sciences* 273 (1597): 2063–2068.

Brainard, M. S., and A. J. Doupe. 2002. What Songbirds Teach Us about Learning. *Nature* 417 (6886): 351–358.

Brower, L. P., W. N. Ryerson, L. L. Coppinger, and S. C. Glazier. 1968. Ecological Chemistry and the Palatability Spectrum. *Science* 161 (3848): 1349–1350. Lincoln Brower, his wife Jane, and several colleagues and students have produced many research papers on this and related subjects.

Campenhausen, M., and H. Wagner. 2006. Influence of the Facial Ruff on the Sound-receiving Characteristics of the Barn Owl's Ears. *Journal of Comparative Physiology A* 192 (10): 1073–1082.

Carvalho, L. S., B. Knott, M. L. Berg, A. T. D. Bennett, and D. M. Hunt. 2010. Ultraviolet-sensitive Vision in Long-lived Birds. *Proceedings of the Royal Society B: Biological Sciences* 278 (1702): 107–114.

Castro, I., Susan J. C., A. C. Gsell, K. Jaffe, A. Cabrera, and C. Liendo. 2010. Olfaction in Birds: A Closer Look at the Kiwi (Apterygidae). *Journal of Avian Biology* 41 (3): 213–218.

Cornell Lab of Ornithology, Comprehensive online guide to North American birds. Includes bird guide with identification, life history, sound, and more. birds.cornell.edu.

Crescitelli, Frederick. 1977. The Avian Eye and Its Adaptations. In *The Visual System in Vertebrates*. Berlin: Springer-Verlag. 550–602.

Cunningham, S. J., J. R. Corfield, A. N. Iwaniuk, I. Castro, M. R. Alley, T. R. Birkhead, and S. Parsons. 2013. The Anatomy of the Bill Tip of Kiwi and Associated Somatosensory Regions of the Brain: Comparisons with Shorebirds. Ed. Melissa J. Coleman. *PLoS ONE* 8 (11): E80036.

Davis, J. K., J. J. Lowman, P. J. Thomas, B. F. H. Ten Hallers, M. Koriabine, L. Y. Huynh, D. L. Maney, P. J. De Jong, C. L. Martin, and J. W. Thomas. 2010. Evolution of a Bitter Taste Receptor Gene Cluster in a New World Sparrow. *Genome Biology and Evolution* 2: 358–370.

Douglas H. D. 2008. Prenuptial Perfume: Alloanointing in the Social Rituals of the Crested Auklet (*Aethia cristatella*) and the Transfer of Arthropod Deterrents. *Die Naturwissenschaften* 95: 45–53. An interesting paper in a fairly unexplored area of ornithology.

Eaton, M. D. 2007. Avian Visual Perspective on Plumage Coloration Confirms Rarity of Sexually Monomorphic North American Passerines. *The Auk* 124 (1): 155–161.

Emery, N. J. 2006. Cognitive Ornithology: The Evolution of Avian Intelligence. *Philosophical Transactions of the Royal Society B: Biological Sciences* 361 (1465): 23–43.

Farner, D. S., and J. R. King. 1973. *Avian Biology*. Vol. 3. New York: Academic Press. Written by international experts from many disciplines, this multi-volume treatise is a comprehensive survey of the established data and principles of avian biology, although a bit dated.

Foster, Mercedes S. 1977. Ecological and Nutritional Effects of Food Scarcity on a Tropical Frugivorous Bird and Its Fruit Source. *Ecology* 58 (1): 73–85.

Gillis, A. M. 1990. What Are Birds Hearing? *BioScience* 40 (11): 810.

Hall, M. I., and C. F. Ross. 2007. Eye Shape and Activity Pattern in Birds. *Journal of Zoology* 271 (4): 437–444.

Hanson, T. 2011. *Feathers*. Philadelphia: Basic Books.

Journey, L., J. P. Drury, M. Haymer, K. Rose, and D. T. Blumstein. 2013. Vivid Birds Respond More to Acoustic Signals of Predators. *Behavioral Ecology and Sociobiology* 67 (8): 1285–1293. Pretty enlightening idea posed here.

Levey, D. J. 1987. Sugar-Tasting Ability and Fruit Selection in Tropical Fruit-Eating Birds. *The Auk* 104: 173–179.

Lipske, M. By a Nose: Birds' Surprising Sense of Smell Biologists Are Discovering That Many Avian Species Rely on Scent for Feeding, Breeding and Other Behaviors. *States News Service*. 29 July 2013. nwf.org/news-and-magazines/national-wildlife/birds/archives/2013/bird-smell.aspx.

Lisney, T. J., D. Rubene, J. Rózsa, H. Løvlie, O. Håstad, and A. Ödeen. 2011. Behavioural Assessment of Flicker Fusion Frequency in Chicken *Gallus gallus domesticus*. *Vision Research* 51 (12): 1324–1332.

Magrath, R. D., B. J. Pitcher, and J. L. Gardner. 2007. A Mutual Understanding? Interspecific Responses by Birds to Each Other's Aerial Alarm Calls. *Behavioral Ecology* 18 (5): 944–951.

Malakoff, David. 1999. Following the Scent of Avian Olfaction. *Science* 286 (5440): 704–705.

Martin, G. R. 2009. What Is Binocular Vision For? A Birds' Eye View. *Journal of Vision* 9 (11): 14.

Morton, Eugene S. 1975. Ecological Sources of Selection on Avian Sounds. *The American Naturalist* 109 (965): 17–34. Pretty classic study on the distance bird sounds carry.

Mouritsen, H., G. Feenders, M. Liedvogel, and W. Kropp. 2004. Migratory Birds Use Head Scans to Detect the Direction of the Earth's Magnetic Field. *Current Biology* 14 (21): 1946–1949.

Nevitt, G. A., and F. Bonadonna. 2005. Sensitivity to Dimethyl Sulphide Suggests a Mechanism for Olfactory Navigation by Seabirds. *Biology Letters* 1 (3): 303–305.

Prum, R. 1999. Development and Evolutionary Origin of Feathers. *Journal of Experimental Zoology* 285: 291–306.

Sayles, I. 1888. Notes Relative to the Sense of Smell in the Turkey Buzzard (*Cathartes aura*). *The Auk* 5 (3): 248–251. Interesting old piece of writing.

Shen, J. X., and Z. M Xu. 1994. Response Characteristics of Herbst Corpuscles in the Interosseous Region of the Pigeon's Hind Limb. *Journal of Comparative Physiology A* 175 (5): 667–674.

Skelhorn, J., and C. Rowe. 2010. Birds Learn to Use Distastefulness as a Signal of Toxicity. *Proceedings of the Royal Society B: Biological Sciences* 277 (1688): 1729–1734.

Sorensen, A. E. 1981. Interactions between Birds and Fruit in a Temperate Woodland. *Oecologia* 50 (2): 242–249.

Sorensen, A. E. 1983. Taste Aversion and Frugivore Preference. *Oecologia* 56 (1): 117–120.

Sturkie's Avian Physiology. 2014. 5th ed. Ed. G. Causey Whittow. New York: Academic Press. A classic that has been around for a long while.

Thorpe, W. H. 1951. The Learning Abilities of Birds. *The Auk* 68 (4): 533–535. A nice overview of what was known at the time of publication.

Waldvogel, J. A. 1990. The Bird's Eye View. *American Scientist* 78: 342–353.

Wetmore, A. 1965. The Role of Olfaction in Food Location by the Turkey Vulture (*Cathartes aura*) Kenneth E. Stager. *The Auk* 82 (4): 661–662.

Whittaker, D., J. Atwell, and E. Ketterson. 2010. Songbird Chemosignals: Volatile Compounds in Preen Gland Secretions Vary among Individuals, Sexes, and Populations. *Behavioral Ecology* 21: 608–614.

Williams, T. D. 1995. *The Penguins*. Oxford: Oxford University Press. One of a series of books on bird families of the world.

Wiltschko, W., U. Munro, H. Ford, and R. Wiltschko. 2006. Bird Navigation: What Type of Information Does the Magnetite-based Receptor Provide? *Proceedings of the Royal Society B: Biological Sciences* 273 (1603): 2815–2820.

Worthington, A. H. 1989. Adaptations for Avian Frugivory: Assimilation Efficiency and Gut Transit Time of *Manacus vitellinus* and *Pipra mentalis*. *Oecologia* 80, no. 3: 381–389.

UPS AND DOWNS

Bicudo, J. E., P. W Eduardo, W. A Buttemer, M. A. Chappell, J. T. Pearson, and C. Bech. 2010. *Ecological and Environmental Physiology of Birds*. Oxford: Oxford University Press. An authoritative series for the experienced reader.

Biewener, A. A. 2011. Muscle Function in Avian Flight: Achieving Power and Control. *Philosophical Transactions of the Royal Society B: Biological Sciences* 366 (1570): 1496–1506.

Birkhead, T. R, J. Wimpenny, and B. Montgomerie. 2014. *Ten Thousand Birds: Ornithology Since Darwin*. Princeton, NJ: Princeton University Press. An interesting overview of the history of ornithological thought.

Calder, W. 1968. Respiratory and Heart Rates of Birds. *The Condor* 70 (4): 358–365.

Chatterjee, S., R. J. Templin, and K. E. Campbell. 2007. The Aerodynamics of *Argentavis*, the World's Largest Flying Bird from the Miocene of Argentina. *Proceedings of the National Academy of Sciences* 104 (30): 12398–12403.

Clark, C. J., and R. Dudley. 2009. Flight Costs of Long, Sexually Selected Tails in Hummingbirds. *Proceedings of the Royal Society B: Biological Sciences* 276 (1664): 2109–2115.

Cline, M. 1905. *Principles of Bird Flight*. New York Academy of Sciences. A serious but totally misinformed treatise on how birds fly; fun to read.

da Vinci, Leonardo. 1938. "Flight," *The Notebooks of Leonardo da Vinci*. Vol. 1. Trans. E. MacCurdy. 471.

Dumont, E. R. 2010. Bone Density and the Lightweight Skeletons of Birds. *Proceedings of the Royal Society B: Biological Sciences* 277: 2193–2198.

Ehrlich, P. R, D. S. Dobkin, and D. Wheye. 1988. *The Birder's Handbook: A Field Guide to the Natural History of North American Birds: Including All Species That Regularly Breed North of Mexico*. New York: Simon & Schuster. A semi-classic of avian natural history information.

Evans, T. R., and L. C. Drikamer. 1994. Flight Speeds of Birds Determined Using Doppler Radar. *The Wilson Bulletin* 106 (1): 154–156.

Gill, Frank B. 2007. *Ornithology*. 3rd ed. New York: W.H. Freeman. Getting to be a standard book for ornithology classes.

Hainsworth, R., and L. Wolf. May 1993. Hummingbird Feeding. *Wildbird Magazine*. A nice general article on hummingbird feeding.

Hanson. *Feathers*. 2011. Philadelphia: Basic Books.

Hartman, F. A. 1955. Heart Weight in Birds. *The Condor* 57 (4): 221–238.

Hedenstron, A., and T. Alerstam. 1995. Optimal Flight Speeds of Birds. *Philosophical Transactions of the Royal Society B: Biological Science* 348 (1326): 471–487.

Jackson, B, Segre, P. and Dial, K. 2009. Precocial Development of Locomotor Performance in a Ground-Dwelling Bird (*Alectoris chukar*): Negotiating a Three-Dimensional Terrestrial Environment. *Proceedings of the Royal Society B: Biological Sciences* 276 (1672): 3457–3466. This research opened a new avenue of thinking about the evolution of flight.

Johannson, L. C., and U. M. L. Norberg. 2000. Biomechanics: Asymmetric Toes Aid Underwater Swimming. *Nature* 407: 982–983.

Kaplan, M. 2014. Why Penguins Cannot Fly. *Nature*: doi:10.1038/nature.2013.13024.

Lissaman, P. B. S., and C. A. Shollenberger. 1970. Formation Flight of Birds. *Science, New Series* 168 (3934): 1003–1005.

Mcnab, B. K. 1994. Energy Conservation and the Evolution of Flightlessness in Birds. *The American Naturalist* 144 (4): 628–642.

Milsom, W. K. 2011. Cardiorespiratory Support of Avian Flight. *The Journal of Experimental Biology* 214 (24): 4071–4072.

Mouillard, L. P. 1881. *The Empire of the Air*. Paris: G. Masson.

Park, J. K., M. Rosen, and A. Hedenstrom. 201. Flight Kinematics of the Barn Swallow (*Hirundo rustica*) Over a Wide Range of Seeds in a Wind Tunnel. *The Journal of Experimental Biology* 204: 2741–2750.

Pennisi, E. 2003. Uphill Dash May Have Led to Flight. *Science, New Series* 299 (5605): 329.

Pennisi, E. 2011. Going the Distance. *Science* 331 (6016): 395–397.

Pennycuick, C. J. 2008. *Modeling the Flying Bird*. London: Elsevier Press, Theoretical Ecology Series. Along with Tucker, perhaps the most famous bird flight researcher.

Prum, R. O., and A. H. Brush. 2002. The Evolutionary Origin and Diversification Of Feathers. *The Quarterly Review of Biology* 77 (3): 261–295. Prum and colleagues changed the way of thinking about how feathers evolved.

Sachs G, J. Traugott, A. P. Nesterova, G. Dell'Omo, and F. Kümmeth. 2012. Flying at No Mechanical Energy Cost: Disclosing the Secret of Wandering Albatrosses. *PLoS ONE* 7(9): e41449.

Speakman, J. R. 2001. The Evolution of Flight and Echolocation in Bats: Another Leap in the Dark. *Mammal Review* 31: 111–130.

Thomas, A. L. R., and A. Balmford. 1995. How Natural Selection Shapes Birds' Tails. *The American Naturalist* 146 (6): 848–868.

Tucker, V. A. 1968. Respiratory Exchange and Evaporative Water Loss in the Flying Budgerigar. *The Journal of Experimental Biology* 48: 67–87.

Tucker, V. A. 1993. Gliding Birds: Reduction of Induced Drag by Wing Tip Slots Between the Primary Feathers. *The Journal of Experimental Biology* 180: 285–310. Tucker has been a well-known researcher for many decades in the field of bird flight.

Videler, John J. 2005 *Avian Flight*. Oxford: Oxford University Press. Some information presented here that is not seen very often.

Weimerskirch, H., J. Martin, Y. Clerquin, P. Alexandre, and S. Jiraskova. 2001. Energy Saving in Flight Formation. *Nature* 413: 697–698.

Wald, C. 2014. Migrating Birds Use Precise Flight Formations to Maximize

Energy Efficiency *Scientific American*. scientificamerican.com/article/migrating-birds-use-preci/.

Xu, X., Z. Zhou, X. Wang, X. Kuang, F. Zhang, and X. Du. 2003. Four-Winged Dinosaurs from China. *Nature* 421 (6921): 335–340. A paper destined to be a classic.

Young, G. F., L. Scardovi, A. Cavagna, I. Giardina, and N. E. Leonard. 2013. Starling Flock Networks Manage Uncertainty in Consensus at Low Cost. Ed. Carl T. Bergstrom. *PLoS Computational Biology* 9 (1): E1002894.

TRAVEL HITHER AND YON

Able, K. P. 1999. *Gatherings of Angels: Migrating Birds and Their Ecology*. Ithaca, NY: Cornell University Press.

Alerstam, T., and A. Hedenstrom. 1998. The Development of Bird Migration Theory. *Journal of Avian Biology* 29 (4): 343.

Beason, R. C. 2005. Mechanisms of Magnetic Orientation in Birds. *Integrative and Comparative Biology* 45: 565–573.

Bell, C. P. 2000. Process in the Evolution of Bird Migration and Pattern in Avian Ecogeography. *Journal of Avian Biology* 31: 258–265.

Berthold, P. 2001. *Bird Migration: A General Survey*. New York: Oxford University Press.

Berthold, P., A. J. Helbig, G. Mohr, and U. Querner. 1992. Rapid Microevolution of Migratory Behaviour in a Wild Bird Species. *Nature* 360 (6405): 668–670.

Birkhead, T. R., J. Wimpenny, and R. D. Montgomerie. 2014. *Ten Thousand Birds: Ornithology since Darwin*. Princeton, NJ: Princeton University Press. A fascinating history of the science.

Boyle, W. A. and C. J. Conway. 2007. Why Migrate? A Test of the Evolutionary Precursor Hypothesis. *The American Naturalist* 169 (3): 344–359.

Calder, C. 1956. Uses of Marking Animals in Ecological Studies: Marking Birds for Scientific Purposes. *Ecology* 37 (4): 665–689. Good review for the time.

Cullen, J. M. 1957. Plumage, Age and Mortality in the Arctic Tern. *Bird Study* 4 (4): 197–207.

Elphick, J. 1995. *The Atlas of Bird Migration: Tracing the Great Journeys of the World's Birds*. New York: Random House. A nicely done and well-illustrated book.

Egevang, C., I. J. Stenhouse, R. A. Phillips, A. Petersen, J. W. Fox, J. R. D. Silt, and C. Cassady. 2010. Tracking of Arctic Terns Sterna paradise Reveals Longest Animal Migration. *Proceedings of the National Academy of Sciences of the United States of America* 107: 2078–2081.

Emlen, S. T., and J. T. Emlen. 1966. A Technique for Recording Migratory Orientation of Captive Birds. *The Auk* 83 (3): 361–367. An explanation of the elegant Emlen Funnel.

Gibbon, E. 1787. *The History Of The Decline And Fall Of The Roman Empire*. Basil: Tourneisen.

Fraser, K. C., T. K. Kyser, and L. M. Ratcliffe. 2008. Detecting Altitudinal Migration Events in Neotropical Birds Using Stable Isotopes. *Biotropica* 40: 269–272.

Griffin, D. R. 1944. The Sensory Basis of Bird Navigation. *The Quarterly Review of Biology* 19 (1): 15–31. A classic paper that reviews what was known at the time.

Gwinner, E. 2003. Circannual Rhythms in Birds. *Current Opinion in Neurobiology* 13(6): 770–778.

Hagen, H. 1975. Beobachtung eines Pfeilstorches in Ost-Afrika, White Stork (*Ciconia ciconia*) with arrow protruding from its body seen in East Africa. *Ornithologische Mitteilungen* 27 (5): 111–112. A very interesting story.

Hansford, D. 2007. Alaska Bird Makes Longest Nonstop Flight Ever Measured. *National Geographic News*, 14 September.

Holland, R. A. 2003 The Role of Visual Landmarks in the Avian Familiar Area Map. *The Journal of Experimental Biology* 206 (11): 1773–1778.

Ingersoll, E. 1968. *Birds in Legend, Fable, and Folklore*. Detroit: Singing Tree Press.

Klassen, M. 1996. Metabolic Constraints on Long-distance Migration in Birds. *The Journal of Experimental Biology* 199: 57–64.

Lack, D. 1968. Bird Migration and Natural Selection. *Oikos* 19 (1): 1–9. David Lack was a pioneer in the field of living bird studies; his book on Darwin's finches, among other works, is a classic.

Newton, I., and K. Brockie. 2008. *The Migration Ecology of Birds*. Amsterdam: Academic.

Lindstrom, A. 1991. Maximum Fat Deposition in Migrating Birds. *Ornis Scandinavica* 22: 12–19.

Marra, P. P., C. M. Francis, R. S. Mulvihill, and F. R. Moore. 2005. The Influence of Climate on the Timing and Rate of Spring Bird Migration. *Oecologia* 142 (2): 307–15.

Nevitt, G. A., and F. Bonadonna. 2005. Sensitivity to Dimethyl Sulphide Suggests a Mechanism for Olfactory Navigation by Seabirds. *Biology Letters* 1 (3): 303–305.

Nilsson, C., R. H. G. Klaassen, and T. Alerstam. 2013. Differences in Speed and Duration of Bird Migration between Spring and Autumn. *The American Naturalist* 181 (6): 837–845.

Robinson, W. D., M. S. Bowlin, I. Bisson, J. Shamoun-Baranes, K. Thorup, R. H. Diehl, T. H. Kunz, S. Mabey, and D. W. Winkler. 2010. Integrating Concepts and Technologies to Advance the Study of Bird Migration. *Frontiers in Ecology and the Environment* 8 (7): 354–361. A recent review of the techniques for studying bird migration.

Rubenstein, D. R., C. P. Chamberlain, R. T. Holmes, M. P. Ayres, J. R. Waldbauer, G. R. Graves, and N. C. Tuross. 2002. Linking Breeding and Wintering Ranges of a Migratory Songbird Using Stable Isotopes. *Science, New Series* 295 (5557): 1062–1065.

Schmidt-Koenig, K., and W. Keeton. 1977. Sun Compass Utilization by Pigeons Wearing Frosted Contact Lenses. *The Auk* 94: 143–145. Another classic experiment.

Suarez, R. K., G. S. Brown, and P. W. Hochhachka. 1986. Metabolic Sources of Energy for Hummingbird Flight. *American Journal of Physiology* 251: 537–543.

Winger, B. M., I. J. Lovette, and D. W. Winkler. 2011 Ancestry and Evolution of Seasonal Migration in the Parulidae. *Proceedings of the Royal Society B: Biological Sciences* 279 (1728): 610–618.

Weidensaul, S. 1999. *Living on the Wind: Across the Hemisphere with Migratory Birds.* New York: North Point. An excellent and fascinating read on the subject.

Wiltschko, W., and R. Wiltschko. 1972. Magnetic Compass of European Robins. *Science, New Series* 176 (4030): 62–64. A pioneering study.

WEATHER SURVIVAL STRATEGIES

Arad, Z., and M. H. Bernstein. 1988. Temperature Regulation in Turkey Vultures. *The Condor* 90 (4): 913–919.

Ashton, K. G. 2002. Patterns of Within-species Body Size Variation of Birds: Strong Evidence for Bergmann's Rule. *Global Ecology and Biogeography* 11 (6): 505–523.

Bartholomew, W. R., C. Lasiewski, and E. C. Crawford Jr. 1968. Patterns of Panting and Gular Fluttering in Cormorants, Pelicans, Owls, and Doves. *The Condor* 70: 31–34.

Bicudo, J., P. W. Eduardo, W. A Buttemer, M. A. Chappell, J. T. Pearson, and C. Bech. 2010. *Ecological and Environmental Physiology of Birds.* Oxford: Oxford University Press.

Cartar, R. V., and R. I. G. Morrison. 2005. Metabolic Correlates of Leg Length in Breeding Arctic Shorebirds: The Cost of Getting High. *Journal of Biogeography* 32 (3): 377–382.

de Vries, J., and M. R. van Eerden. 1995. Thermal Conductance in Aquatic Birds in Relation to the Degree of Water Contact, Body Mass, and Body Fat: Energetic Implications of Living in a Strong Cooling Environment *Physiological Zoology* 68 (6): 1143–1163.

Duncan R. P., T. M. Blackburn, and D. Sol. 2003. The Ecology of Bird Introductions. *Annual Review of Ecology and Systematics* 34: 71–98.

Dunn, E. H., and D. L. Tessaglia. 1994. Predation of Birds at Feeders in Winter. *Journal of Field Ornithology* 65: 8–16.

Ellis, H. I., and J. R. Jehl. 2003. Temperature Regulation and the Constraints of Climate in the Eared Grebe. *Waterbirds* 26 (3): 275–279.

Hatchwell, B. J., S. P. Sharp, M. Simeoni, and A. Mcgowan. 2009. Factors Influencing

Overnight Loss of Body Mass in the Communal Roosts of a Social Bird. *Functional Ecology* 23 (2): 367–372.

Hilton, G. M., G. D. Ruxton, and W. Cresswell. 1999. Choice of Foraging Area with Respect to Predation Risk in Redshanks: The Effects of Weather and Predator Activity. *Oikos* 87: 295–302.

Hohtola, E. 2004. Shivering Thermogenesis in Birds and Mammals. In *Life in the Cold: Evolution, Mechanisms, Adaptation, and Application*, 241–252. Barnes BM, Carey HV, eds. Twelfth International Hibernation Symposium. Institute of Arctic Biology, University of Alaska, Fairbanks, AK.

Holick, M. F., T. C. Chen, Z. Lu, and E. Sauter. 2007. Vitamin D and Skin Physiology: A D-Lightful Story. *Journal of Bone and Mineral Research* 22 (S2): V28–33.

James, F. C. 1970. Geographic Size Variation in Birds and Its Relationship to Climate. *Ecology* 51 (3): 365–390. A nice overall perspective of the idea.

McKechnie, A. E., and B. G. Lovegrove. 2002. Avian Hypothermic Facultative Responses: A Review. *The Condor* 104: 705–724.

Meiri, S., and T. Dayan. 2003. On the Validity of Bergmann's Rule. *Journal of Biogeography* 30: 331–351. A nice overview of organisms adhering to Bergmann's rule.

Nudds, R. L., and S. A. Oswald. 2007. An Interspecific Test of Allen's Rule: Evolutionary Implications for Endothermic Species. Evolution 61: 2839–2848.

Petit, D. R. 1989. Weather-Dependent Use of Habitat Patches by Wintering Woodland Birds. *Journal of Field Ornithology* 60: 241–247.

Roberts, T. S. 1907. A Lapland Longspur Tragedy: Being an Account of a Great Destruction of These Birds during a Storm in Southwestern Minnesota and Northwestern Iowa in March, 1904. *The Auk* 24: 369–377.

Root, T. 1988. Energy Constraints on Avian Distributions and Abundances. *Ecology* 69: 330-339.

Rowling, J. K. 2003. *Harry Potter and the Order of the Phoenix*. New York: Scholastic, Inc.

Russell, D. G. D., W. J. L. Sladen, and D. G. Ainley. 2012. Dr. George Murray Levick (1876–1956): Unpublished Notes on the Sexual Habits of the Adélie Penguin. *Polar Record* 48 (4): 387–393.

Schleucher, E. 2004. Torpor in Birds: Taxonomy, Energetics, and Ecology. *Physiological and Biochemical Zoology* 77: 942–949. An excellent review of the subject.

Symonds, M. R. E., and G. N. J. Tattersall. 2010. Geographical Variation in Bill Size across Bird Species Provides Evidence for Allen's Rule. *The American Naturalist* 176 (2): 188–197.

Tattersall, G. J., D. V. Andrade, and A. S. Abe. 2009. Heat Exchange from the Toucan Bill Reveals a Controllable Vascular Thermal Radiator. *Science* 325 (5939): 468–470.

Thomas, N. J., B. Hunter., and C. T. Atkinson. 2007. *Infectious Diseases of Wild Birds*. Ames, IA: Blackwell Publishing.

Thomas, R. J., and I. C. Cuthill. 2002. Body Mass Regulation and the Daily Singing Routines of European Robins. *Animal Behaviour* 63 (2): 285–295.

Weathers, W. W., P. J. Hodum, and J. A. Blakesley. 2001. Thermal Ecology and Ecological Energetics of California Spotted Owls. *The Condor* 103: 678–690.

Williams, J. B., and B. I. Tieleman. 2005. Physiological Adaptation in Desert Birds. *BioScience* 55 (5): 416–425.

Wilson, G. R., S. J. Cooper, and J. A. Gessaman. 2004. The Effects of Temperature and Artificial Rain on the Metabolism of American Kestrels (Falco Sparverius). *Comparative Biochemistry and Physiology Part A: Molecular & Integrative Physiology* 139 (3): 389–394.

Wade, J. L. 1966. *What You Should Know About the Purple Martin: America's Most Wanted Bird*. Griggsville, IL: J.L. Wade.

Wolf, B. O., K. M. Wooden, and G. E. Walsberg. 1996. The Use of Thermal Refugia by Two Small Desert Birds. *The Condor* 98: 424–428.

BIRD COMMUNITIES

Ardrey, R. 1967. *The Territorial Imperative: A Personal Inquiry into the Animal Origins of Property and Nations*. London: Collins.

Bent, A. C. 1919–1968. Life Histories of North American Birds. 21 Volumes. U.S. Government Printing Office. *National Museum Bulletins* 107–137.

Brock, T. and B. C. Sheldon. 2010. Individuals and Populations: the Role of Long Term Individual-based Studies of Animals in Ecology and Evolutionary Biology. *Trends in Ecology and Evolution* 25: 562–573.

Brush, A. H., and G. A. Clark. 1983. *Perspectives in Ornithology: Essays Presented for the Centennial of the American Ornithologists' Union*. Cambridge: Cambridge University Press.

Cody, M. L. 1974. *Competition and the Structure of Bird Communities*. Princeton, NJ: Princeton University Press. A well-known piece of research.

Cordeiro, N. and H. F. Howe. 2003. Forest Fragmentation Severs Mutualism Between Seed Dispersers and an Endemic African Tree. *PNAS* 100 (24): 14052–14056.

Dale, V. H., C. M. Crisafulli, and F. J. Swanson. 2005. Ecology: 25 Years of Ecological Change at Mount St. Helens. *Science* 308 (5724): 961–962.

Darwin, C. 1859. *On the Origin of Species by Means of Natural Selection, or the Preservation of Favoured Races in the Struggle for Life*. London: J. Murray. The classic of all classics.

Dhondt, A, A., and R. Eyckerman. 1980. Competition Between the Great Tit and the

Blue Tit Outside the Breeding Season in Field Experiments. *Ecology* 61: 1291–1296.

Frakes, R. A., and R. E. Johnson. 1982. Niche Convergence in Empidonax Flycatchers. *The Condor* 84: 286–291.

Freed, L. A., and R. L. Cann. 2009. Negative Effects of an Introduced Bird Species on Growth and Survival in a Native Bird Community. *Current Biology* 19 (20): 1736–1740.

Gause, G. F. 1932. Experimental studies on the struggle for existence: two species of yeast. *The Journal of Experimental Biology* 9: 389–940. A classic paper.

Geisel, T. S. 1955. *On Beyond Zebra!* New York: Random House. Another classic.

Grinnell, J. 1904. The Origin and Distribution of the Chest-Nut-Backed Chickadee. *The Auk* 21 (3): 364–382.

Grinnell, J. 1917. The Niche-Relationships of the California Thrasher. *The Auk* 34 (4): 427–433.

Hensley, M. M., and J. B. Cope. 1951. Further Data on Removal and Repopulation of the Breeding Birds in a Spruce-Fir Forest Community. *The Auk* 68 (4): 483–493.

Hulme, P., and C. W. Benkman. 2002. Granivory. In *Plant-Animal Interactions: An Evolutionary Approach.* Edited by C. Herrera and O. Pellmyr. New York: Blackwell Scientific Publishing. 132–154.

Hutchinson, G. E. 1959. Homage to Santa Rosalia, Or Why Are There So Many Kinds of Animals? *The American Naturalist* 93: 145–159. A classic paper.

Johnston, D. W., and E. P. Odum. 1956. Breeding Bird Populations in Relation to Plant Succession on the Piedmont of Georgia. *Ecology* 37: 50–62.

Lederer, R. J. 1977. Winter Feeding Territories in the Townsend's Solitaire. *Bird Banding* 48: 11–18.

Lewis, S., T. N. Sherratt, K. C. Hamer, and S. Wanless. 2001. Evidence of Intra-Specific Competition for Food in a Pelagic Seabird. *Nature* 412: 816–819.

MacArthur, R. H. 1958. Population Ecology of Some Warblers of Northeastern Coniferous Forests. *Ecology* 39 (4): 599–619.

MacArthur, R. H., and J. W. MacArthur. 1961. On Bird Species Diversity. *Ecology* 42 (3): 594–598.

Moermond, T. C., and J. S. Denslow. 1985. Neotropical Avian Frugivores: Patterns of Behavior, Morphology, and Nutrition, with Consequences for Fruit Selection. *Ornithological Monographs* 36, *Neotropical Ornithology*: 865–897.

Nolan, V., E. D. Ketterson, and C. F. Thompson. 1997. *Current Ornithology*. New York: Plenum.

Noon, B. R. 1981. The Distribution of an Avian Guild along a Temperate Elevational Gradient: The Importance and Expression of Competition. *Ecological Monographs* 51:105–124.

Nussey, D. H., E. Postma, P. Gienapp, and M. E. Visser. 2005. Selection on Heritable

Phenotypic Plasticity in a Wild Bird Population. *Science, New Series* 310: 304–306.

Palmer, T. S. 1922. Game as a National Resource. Washington, D.C.: U.S. Dept. of Agriculture.

Peterson, R. T. 2008. *Peterson Field Guide to Birds of North America*. Boston: Houghton Mifflin.

Shugart, H. H. Jr., and D. James. 1973. Ecological Succession of Breeding Bird Populations in Northwestern Arkansas. *The Auk* 90: 62–77.

Tarvin, K. A., M, C. Garvin, J. M. Jawor, and K. A. Dayer. 1998. A Field Evaluation of Techniques Used to Estimate Density of Blue Jays. *Journal of Field Ornithology* 69: 209–222.

Terborgh, J. 1977. Bird Species Diversity on an Andean Elevational Gradient. *Ecology* 58: 1007–1019.

Terborgh, J., S. K. Robinson, T. A. Parker III, C. A. Munn, and N. Pierpont. 1990. Structure and Organization of an Amazonian Forest Bird Community. *Ecological Monographs* 60 (2): 213–237.

Thornton, I. W., R. A Zann, P. A. Rawlinson, C. R. Tidemann, A. S. Adikerana, and A. H. Widjoya. 1988. Colonization of the Krakatau Islands by Vertebrates: Equilibrium, Succession, and Possible Delayed Extinction. *Proceedings of the National Academy of Sciences* 85 (2): 515–518.

University of Nebraska Press / University of Nebraska-Lincoln Libraries-Electronic Text Center. *The Journals of the Lewis and Clark Expedition*. lewisandclarkjournals.unl.edu.

Vale, T. R., A. J. Parker, and K. C. Parker. 2010. Bird Communities and Vegetation Structure in the United States. *Annals of the Association of American Geographers* 72: 120–130.

Vall-Llosera, M., and D. Sol. 2009. A Global Risk Assessment for the Success of Bird Introductions. *Journal of Applied Ecology* 46 (4): 787–795.

Weins, J., ed. 1992. *The Ecology of Bird Communities*. Vol 1. *Foundations and Patterns*. Cambridge: Cambridge University Press.

Wiens, J. A., J. T. Rotenberry, and B. Van Horne. 1987. Habitat Occupancy Patterns of North American Shrubsteppe Birds: The Effects of Spatial Scale. *Oikos* 48: 132–147.

Willson, M. F. 1969. Avian Niche Size and Morphological Variation. *The American Naturalist* 103 (933): 531–542.

Willson, M. F. 1974. Avian Community Organization and Habitat Structure. *Ecology* 55 (5): 1017–1029.

HUMAN INFLUENCES

American Bird Conservancy. 2014. Global Warming Threatens Many Bird Species. abcbirds.org/abcprograms/policy/globalwarming/index.html.

Audubon. 2014. Birds and Climate Change: On the Move. climate.audubon.org/.

Blackburn, T. M. 2004. Avian Extinction and Mammalian Introductions on Oceanic Islands. *Science* 305: 1955–1958.

Bonier, F., P. R. Martin, and J. C. Wingfield. 2007. Urban birds have broader environmental tolerance. *Biology Letters* 3 (6), 670–673.

Brockie, B. Introduced Animal pests—Impact of Animals on the Bush. Te Ara—The Encyclopedia of New Zealand, Updated 20 May 2014. TeAra.govt.nz/en/speech/10106/joseph-bankss-journal.

Brown, C. R., and M. B. Brown. 2013. Where Has All the Road Kill Gone? *Current Biology* 23 (6): 233–234.

Coleman, J. S., and S. A Temple. 1995. How Many Birds Do Cats Kill? *Wildlife Control Technology* (July–August): 44.

DeGraaf, R. M., and J. M. Wentworth. 1986. Avian Guild Structure and Habitat Associations in Suburban Bird Communities. *Urban Ecology* 9: 399–412.

Del-hoyo, J., A. Elliot, and J Sardatal. 2002. *Handbook of Birds of the World, Jacamars to Woodpeckers*. Vol. 7. Barcelona: Lynx.

Durães, R., L. Carrasco, T. B. Smith, and J. Karubian. 2013. Effects of Forest Disturbance and Habitat Loss on Avian Communities in a Neotropical Biodiversity Hotspot. *Biological Conservation* 166: 203–11.

Eeva, T. 2009. The Effects of Diet Quality and Quantity on Plumage Colour and Growth of Great Tit Parus Major Nestlings: A Food Manipulation Experiment along a Pollution Gradient. *Journal of Avian Biology* 40 (5): 491–499.

Engels, S., N-L. Schneider, N. Lefeldt, C. M. Hein, M. Zapka, A. Michalik, D. Elbers, A. Kittel, P. J. Hore, and H. Mouritsen. 2014. Anthropogenic Electromagnetic Noise Disrupts Magnetic Compass Orientation in a Migratory Bird. *Nature* 509: 353–356.

Erickson, W. P., G. D. Johnson, and D. P. Young Jr. 2005. A Summary and Comparison of Bird Mortality from Anthropogenic Causes with an Emphasis on Collisions. *USDA Forest Service.*

Evans, K. L., D. E. Chamberlain, B. J. Hatchwell, R. D. Gregory, and K. J. Gaston. 2011. What Makes an Urban Bird? *Global Change Biology* 17 (1): 32–44.

Ralph, C. J., and T. D. Rich, eds. 2002. *Bird Conservation Implementation and Integration in the Americas: Proceedings of the Third International Partners in Flight Conference*. March 20–24; Asilomar, California; Volume 2. *General Technical Report* 191: 1029–1104. U.S. Dept. of Agriculture, Forest Service, Pacific Southwest Research Station.

Evans, K. L., J. Newton, K. J. Gaston, S. P. Sharp, A. McGowan, and B. J. Hatchwell. 2012. Colonisation of Urban Environments is Associated with Reduced Migratory Behaviour, Facilitating Divergence From Ancestral Populations. *Oikos* 121: 634–640.

Gates, J. E., and J. A. Mosher. 1981. A Functional Approach to Estimating Habitat Edge Width for Birds. *American Midland Naturalist* 105: 189–192.

Inger. R., R. Gregory, J. P. Duffy, I. Stott, P. Voříšek, and K. J. Gaston. 2014. Common European Birds Are Declining Rapidly While Less Abundant Species' Numbers Are Rising. *Ecology Letters*. onlinelibrary.wiley.com/doi/10.1111/ele.12387/full.

Kempenaers, B., P. Borgstrom, P. Loës, E. Schlicht, and M. Valcu. 2010. Artificial Night Lighting Affects Dawn Song, Extra-Pair Siring Success, and Lay Date in Songbirds. *Current Biology* 20 (19): 1735–1739.

Klem, D. 2012. Avian Mortality on Windows: the Second Largest Source of Bird Mortality on Earth. *Proceedings of the Fourth International Partners in Flight Conference: Tundra to Tropics*: 244–251.

Legagneux, P., and S. Ducatez. 2013. European Birds Adjust Their Flight Initiation Distance to Road Speed Limits. *Biology Letters* 9 (5): 20130417.

Marsh, George Perkins. 1865. *Man and Nature*. New York: Charles Scribner.

Nijhuis, M. 2014. A Gathering Storm for North American Birds. *Audubon* 116 (5): 24–30.

Quintero, I., and J. J. Wiens. 2013. Rates of Projected Climate Change Dramatically Exceed past Rates of Climatic Niche Evolution among Vertebrate Species Ed. Luke Harmon. *Ecology Letters* 16 (8): 1095–1103.

San Francisco Planning Department. 2011. Standards for Bird-Safe Buildings. sf-planning.org/ftp/files/publications_reports/bird_safe_bldgs/Standards_for_Bird-Safe_Buildings_8-11-11.pdf.

Schmiegelow, F. K. A., and M. Monkkonen. 2002. Habitat Loss and Fragmentation in Dynamic Landscapes: Avian Perspectives From the Boreal Forest. *Ecological Applications* 12: 375–389.

University of Oldenburg. 2014. "Electrosmog" Disrupts Orientation in Migratory Birds, Scientists Show. *ScienceDaily*, 8 May 2014.

Wilson, A., and T. M. Brewer. 1840. *American Ornithology*. Boston: Otis, Broaders.

Wood, W. E., and S. M. Yezerinac. 2006. Song Sparrow (*Melospiza melodia*) Song Varies With Urban Noise. *The Auk* 123: 650–659.

ACKNOWLEDGMENTS

In the research for this book I spent many hours perusing the voluminous scientific literature, which seems to be expanding at an exponential rate, as well as information in the popular literature, and on the worldwide web from sources I have learned to trust. California State University, Chico, provided access to its brick and mortar library and ingress into the vast world of scientific journals in its stacks and online.

Whatever I found invariably led to even more inquiries, so I was happy to be able to rely on friends and colleagues to provide comments on both my information and writing style. Tim Ruckle, a former NASA editor and avid birdwatcher, commented on the manuscript with a discerning eye. I am especially grateful to Steve King, an active Audubon member, who not only gave my drafts a close edit, but helpfully pointed out areas where I was not sufficiently lucid.

Juree Sondker of Timber Press was in at the initial brainstorming and shepherding the book through the approval process. She kept me on track with insightful comments, edits, and technical advice. The art department at Timber Press oversaw the art and illustrations.

My wife, Carol Burr, drew some of the artwork and was supportive and encouraging all along, putting up with my frequent disappearance into the bowels of our home office to stare at an LCD screen much of the day. Her long tenure as an English professor was also handy in helping me craft my prose.

And enormous thanks to copyeditor Mollie Firestone who inspected and corrected my prose with an artful eye. Zeroing in on detail, parsing sentences, and asking for explanations kept me on my authorial toes.

PHOTO AND ILLUSTRATION CREDITS

Anna Eshelman, based on a diagram by the author, page 148

Anna Eshelman, based on a diagram by Jimfbleak, edited by McSush, public domain on Wikimedia Commons, page 53

Anna Eshelman, based on a diagram by L. Shyamal, public domain on Wikimedia Commons, page 57

Anna Eshelman, based on the map "Biological Flyways" by Michael A. Johnson, North Dakota Game and Fish, U.S. Fish and Wildlife Service, page 141

Carol Burr, pages 29, 36

Dario Sanches, page 235

Donald R. Miller Photography, page 30

Kate Francis, based on a drawing by Carol Burr, page 188

Kate Francis, based on a drawing by Ekann, used under a Creative Commons Attribution—ShareAlike 4.0 International license, Wikimedia Commons, page 162

Kate Francis, based on Emlen and Emlen 1966, pages 363–364, and on a drawing by L. Shyamal, public domain on Wikimedia Commons, pages 121–122

Kate Francis, based on a drawing by Lucas Martin Frey, used under a Creative Commons Attribution 3.0 Unported license, Wikimedia Commons, pages 180–181

Kate Francis, based on a drawing by U.S. National Oceanic and Atmospheric Administration, National Climatic Data Center, public domain on Wikimedia Commons, page 219

Kerry Cesen, based on drawings by Carol Burr, pages 32, 137, 189

Kerry Cesen, based on a drawing by Conty, public domain on Wikimedia Commons, page 100

FLICKR

Used under a Creative Commons Attribution—ShareAlike 2.0 Generic License

Benson Kua, page 222

brewbooks, page 72

Dan Pancamo, page 42

Dominic Sherony, page 168
Greg Miles, page 213
Jinx!, page 232
Juan Emilio, page 228
Kate/Carine06, page 37
Mike's Birds, page 68
Shanthanu Bhardwaj, page 154

Used under a Creative Commons Attribution 2.0 Generic License
Angelo DeSantis, page 132
Denali National Park and Preserve, photo by Tim Rains, page 98
D. Faulder, page 190 right
John Benson, page 107
Kevin Cole, page 130
Matt MacGillivray, page 159
Mike Baird, page 190 left
U.S. Fish and Wildlife Service, photo by Mike Morel, page 40
William Warby, page 152

IPERNITY

Used under a Creative Commons Attribution 3.0 Unported license
Ian Kirk, page 126

PUBLIC DOMAIN IMAGES

U.S. Fish and Wildlife Service, photo by Dave Menke, pages 103, 147
U.S. Fish and Wildlife Service, photo by George Gentry, page 233

WIKIMEDIA

Used under a Creative Commons Attribution—ShareAlike 4.0 International license
Avitopia, page 155
Bartosz Kosiorek, page 94
Kokopelado, page 239
Ryan O'Donnell, page 110

Used under a Creative Commons Attribution—ShareAlike 3.0 Unported license
4028mdk09, page 160
Akshay Charegaonkar, page 197
Cephas, pages 14, 65, 192

Chris Harshaw, page 59

Chuck Szmurlo, page 61

Dori, page 49

Dick Daniels, pages 51, 193, 218, 238

Dr. Tejinder Singh Rawal, page 114

Francis C. Franklin, page 186

Hugo Pedel, page 104

H. Raab, page 19

Jerry Friedman, page 27

Jose Lopez, Jr., edited by Struthious Bandersnatch, page 164

JJ Harrison, page 70

Martin Wegmann, page 216

Mdf, pages 2, 34

Olaf Oliviero Riemer, page 165

Rick elis.simpson, page 76

Shao, page 179

Symphony999, page 135

Umesh Srinivasan, page 117

Шатилло Г.В. (Shatilla G.V.), page 169

Yathin S. Krishnappa, page 85

Used under a Creative Commons Attribution—ShareAlike 2.5 Generic license

Thermos, pages 45, 143

Andreas Trepte (http://foto.andreas-trepte.de), pages 78, 127

toony & svtiste, page 87

L. Shyamal, page 97

Used under a Creative Commons Attribution-ShareAlike 2.0 Generic license

Fernando Flores, page 227

Used under a Creative Commons Attribution 2.0 Generic license

Angelo DeSantis, page 133

Tim Sträter, page 202

Public domain on Wikimedia Commons

Charles Joseph Hullmandel, page 224

David Helton and WMMS in Exit Magazine, page 119

Jacob Peter Gowy, page 83

Jason Corriveau, page 90

John Gould, page 21

John Gould & Henry Constantine Richter, page 125

Muriel Gottrop, page 95

Mysid, page 64

Parker & Coward, page 183

Philip Henry Gosse, page 31

U.S. Fish and Wildlife Service, photo by Art Sowles, page 108

U.S. Fish and Wildlife Service, National Image Library, photo by Dave Menke, page 23

Wenceslaus Hollar via the University of Toronto Thomas Fisher Rare Book Library, page 204

INDEX